Torquil Barker

concepts in practice
lighting

Lighting Design in Architecture

Acknowledgements

Erco Lighting Limited
38 Dover Street, London W1X 3RB, England
Tel 0171 408 0320 Fax 0171 409 1530

Maurice Brill Lighting Design Ltd.
3rd Floor, 3 Swan Field Court, 48 Chilton Street, London E2 6D, England
Tel 0171 613 0456 Fax 0171 613 0465

John Bradley Associates, Mechanical and Electrical Engineering Consultants,
Studio 4, Pickfords Wharf, Clink Street, London SE1 9DC, England
Tel 0171 407 6616 Fax 0171 407 6592

LEC Lyon, 6 Rue Fornet, F-9006 Lyon, France
Tel 3378 529792 Fax 3378 246056

Linda Ferry, Architectural Illumination, PO Box 22690, Monterey, California, USA
Tel (408) 649 3711 Fax (408) 375 5897

Professor Wolfgang Doring, Architect, Dusseldorfer Strasse, 155,
40545 Dusseldorf, Germany
Tel 0211 559 0010 Fax 0211 559 0012

Behnisch & Partener, Breie Architekten BDA, Buro Sillenbuch,
Corch-Fock-Strausse 30, 70619 Stuttgart, Germany
Tel 0711 47656-0 Fax 0711 47656-56

Riken Yamamoto & Field Shop, Urban-Yokohama, BLD-7F,
1-5-14 Kitasaiwai Nishiku, Yokohama, Japan 220
Tel 045 323 6010 Fax 045 323 6012

For Mary and George Bell Barker

First published in Great Britain 1997

© Torquil Barker 1997

All rights reserved. No part of this publication
may be reproduced in any form or by any means,
without permission from the publisher
A CIP catalogue record for this book is available
from the British Library

ISBN 0 7134 7876 4

Printed in Singapore

For the Publishers

B.T. Batsford Ltd
583 Fulham Road
London SW6 5BY

Contents

Introduction	4
1: Daylight	18
2: Retail	30
3: Hotels	40
4: Art galleries	58
5: Exteriors	74
6: Residential	90
7: Past and future predictions	110
8: Appendix	120
Glossary	125
Bibliography	127
Index	128

Introduction

Light has been worshipped, converted to power, used to measure time, to propagate plants, guide ships and cut through sheets of steel. Without light the human race could not exist. Light is also the agency by which objects are rendered visible. To work with light is to handle a both dangerous and beneficial medium. Light can blind the eye, while laser surgery can restore lost sight. Light can be used as a cure for depression or as a tool of sensory deprivation. Since the beginning of time we have had a dynamic relationship with the phenomenon of light.

Since the discovery of fire human beings have controlled an element which has allowed them to see in darkness, kept them warm, could be used to cook meat, and helped to keep wild beasts at bay. They also discovered fire would burn them, if not handled with respect. In much the same way that electricity can end or preserve life, so too fire proved a double-edged sword.

Early lighting design

The first construction to utilize light in a creative and spectacular manner was the Hagia Sofia, a church built between 532 and 537 AD in Istanbul, during the reign of the Emperor Justinian, a ruler obsessed with religion and war. Hagia Sophia, built as a physical manifestation of the heavenly realms above, still stands as a triumph of beauty and design. Its ascending domes are pierced by shafts of light, causing the interior to glow with a burnished splendour as if the very walls had been lit from within. A mysterious element was brought into being which was more than the sum of light meeting architecture. From the shimmer of mosaics and marbles there is a transmission of the mysticism woven within the fabric of Eastern Christianity.

It was not until the Gothic cathedrals of the thirteenth and fourteenth century were constructed that such a transcendent clarity was once more captured in the geometric equation of low northern light and coloured glass. The first of these cathedrals to display the beauty of luminous windows was built by Abbot Suger of Saint Denis in France, around 1140 AD. The precision and skill required to construct this building was extraordinary and the result was essentially emotional. Contemporary accounts relate that it was the richness and scale of the windows that stirred the people's imagination. These stained glass windows formed a translucent gateway to worship.

Figure 1
Church of San Lorenzo, Florence, Italy. A sequential layering of illumination leads the visitor towards the main altar. Erco Lighting Limited.

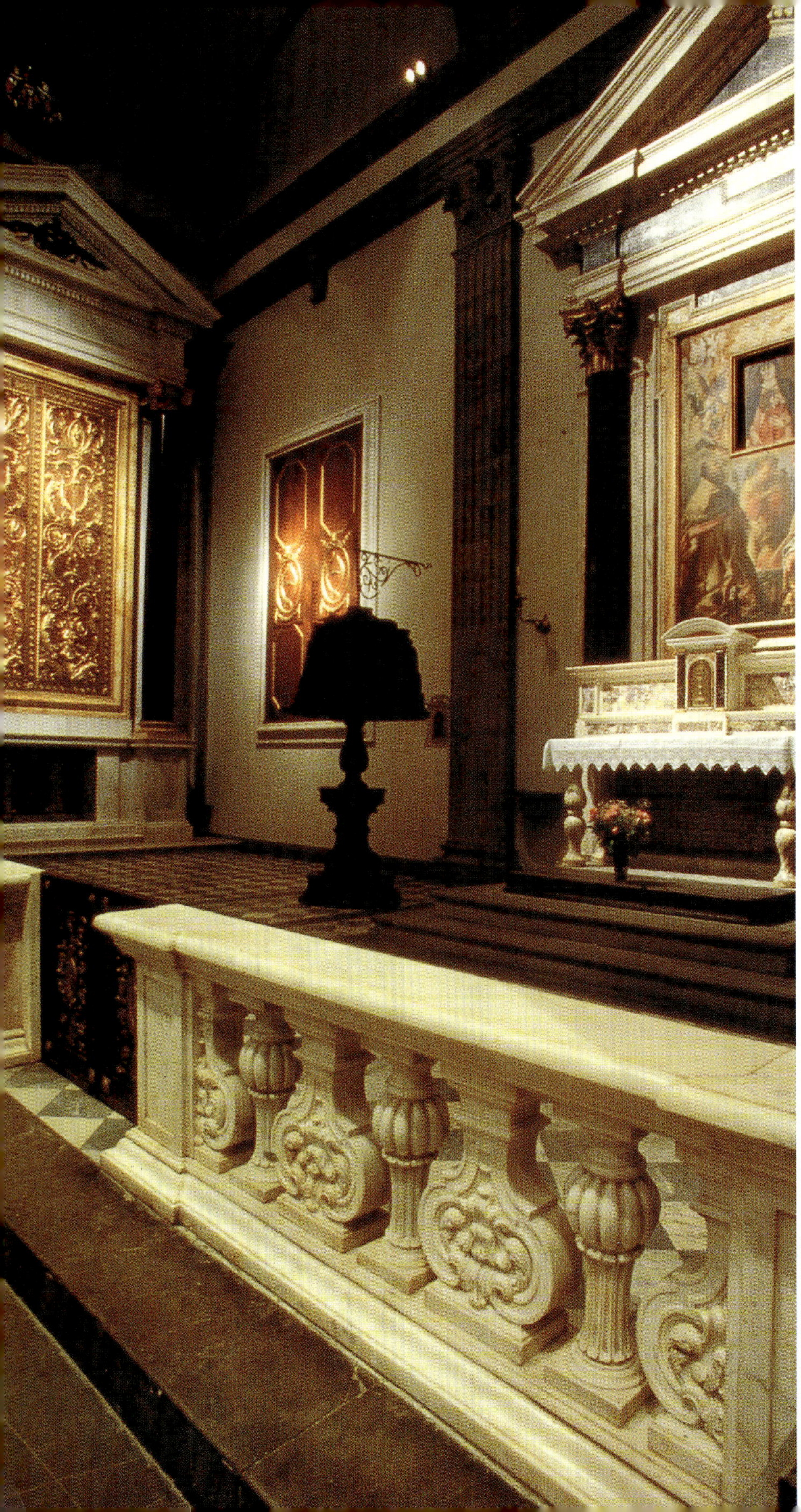

Figure 2
Church of San Lorenzo, Florence, Italy. Brunelleschi's architecture revealed by hidden low voltage halogen spotlights. Erco Lighting Limited.

They filled the church with a soft light that suggested a world beyond harsh reality. These windows were a rich synthesis of geometry and colour, imposing yet also reassuring in their grandeur.

Stained glass had been extant since the fourth century and by the twelfth century was widely used to glorify the state, Church and monarch. The effect of these translucent tableaux, revealing the realms of saints and angels, was so persuasive that Gothic architecture was adapted to accommodate larger windows, finding expression in the cathedrals of Amiens, Reims, and Chartres, where magnificent stained glass served as a bridge between physical and spiritual worlds. The tonal quality of the windows altered throughout the day, depending on the elevation of the sun and climatic conditions, so that there was a subtle sense of change from dawn to dusk.

The craftworkers who created these windows wished to bring to the congregation an idea of what the heavenly world might be like. At this time, the Church was very powerful and its buildings were constructed as a form of proof in the strength of belief. It was the first time in centuries that light had been used with a fluency and beauty that rivalled and at times excelled the architecture it revealed.

It was not until the Italian Renaissance that painting began to challenge the beauty of stained glass. Painting is not design with light as such, but it was the first time a study of light was transferred to a flat surface. Artists found that light and shade could create an illusion of form.

Leonardo sometimes painted under a gauze tent, thus softening the shadows on the model's face. Light and shade were manipulated within large paintings, to create form and movement. Artists such as Botticelli, Raphael and Michelangelo used their knowledge of light and dark to suggest harmony, tension and counterpoint, within a rhythmic unified composition. There is an intensity to great painting that evokes an emotional reaction beyond language. Fine architectural spaces also have this power to deeply affect the observer.

Lighting design in the twentieth century

In the twentieth century, the paintings of the French Impressionists were in part a study of how light reflects or is absorbed by different surfaces. Many of these works were created out of doors, so that the artist shared the light of the subject painted.

In architecture the work of Frank Lloyd Wright, Gropius, Antonio Gaudi and Mies van der Rohe found quite disparate solutions to lighting the interior. It was however, within another arena that electric light challenged the boundaries of the possible.

This was aboard the great ocean liners of the 1930s and 1940s. The *Berengaria*, *Bremen*, *Isle de France* and *Normandie* were suffused with the glow of electrified lamps. Here was a flamboyant theatricism which merged the richness of Byzantia with the ethereal windows of Chartres. The liner *Paris* sported a glass dance floor illuminated from beneath, so that the dancers appeared to float upon a web of colour.

These ships were roving ambassadors, capable of satisfying the most jaded of travellers, whether by seductive ambience or a superbly selective French wine list. Fluted floor and wall uplighters created a translucent architecture through which light flowed. The zenith of marine Art Deco was achieved in the *Normandie* which was a spangled floating palace dedicated to the beauty of illumination. Concealed and indirect light brought a mercurial brilliance to walls of onyx and murals of silver and gold. Within this setting the voyage was conducted as a stylized theatre of manners. A lavish use of fluid space was heightened by illusions of mirror and glass. In the Bauhaus-style nightclub there were Jazz-age arabesques of neon. Everywhere were bold applications of electric light that would not have been acceptable on land. In the dining room, twenty-foot-high light tubes touched the ceiling, where glacial crystal pendants were suspended like stalactites. This room was over three hundred feet in length and could seat up to seven hundred diners.

These ships reflected the design ethos of their country of origin and competition was not just for the coveted 'Blue Riband', but as to who could create the most dazzling interiors. Crossing the Atlantic was an emotional experience, linking old worlds with new. These floating temples of pleasure conjured a sensory paradise far removed from that envisaged by the stained glass craftsmen of Chartres.

The art of lighting design

Rooms in different types of buildings are often described as spacious, ethereal, claustrophobic or depressing. These are words that have an emotional significance. An interior space is seldom defined by a code or number sequence. 'Athena Ergene', the elusive goddess-muse of architecture is not so easily contained within the number-crunched pixels of computer graphics.

It is light that reveals the architecture. There is a basic problem in communication when trying to express in words what is essentially a visual experience. The subtle qualities of light do not fit easily to a verbal or mathematical explanation. Just as some scientists have attempted to rationalize the mysteries of the universe, so the lighting engineer reduces light to neat grids of numbers. Each person's eyes see and experience light differently. No two stage-lighting designers will illuminate one scene in a play the same way. The variables are such that they cannot be quantified.

Before the arrival of the lighting designer within the commercial arena, lighting was provided either by an architect or lighting engineer. The engineer thought in equations, while the architect was concerned with surface and volume. Often neither had time to give lighting sufficient consideration. Lighting was sometimes hooked on the end of a job, and quality would be governed by the amount of money remaining. At first, both these professions were resentful and wary of the lighting designer. With time, most have come to appreciate the lighting designer's grasp of the technical problems of light measurement, and the way it relates to the complex demands of spatial aesthetics.

Many of today's designers have come from a theatrical background which demanded technical understanding of the equipment, along with an artist's consideration of dramatic mood. The stage encourages the designer to envisage in terms of space. Light is a powerful manipulator of space. Lighting engineers do not always consider the psychological impact that light will have on an interior. They can be calculation-led by sets of illuminance levels, and, like carpet fitters, spread light evenly from wall to wall, in a blanket interpretation of the relevant regulations.

Humans do not work well in environments which are blandly and evenly lit. It is not uncommon for office workers to have to exist in a relamped snowscape of white metal halide light, which, although giving good colour rendition, creates a hard, alienating, atmosphere. No wonder that many complain of illness and of headaches, only to be told that it is all in their imagination.

Psychologically, we have adapted to warm light sources: first firelight, candles, paraffin lamps, and finally the electric bulb, which produces a yellow-orange light, perhaps more apparent in photographs than to the naked eye.

Lighting is about people. There is no point specifying a highly energy-efficient light source for a workspace if employees feel uncomfortable working under it. Most people prefer to have slight variations in lighting levels, and when possible, lighting which tends towards warmth. There was for a while the idea that a cold, even, lighting would help to keep a workforce awake and efficient. Though bosses were happy to deploy this light in their employees' workspace, it was interesting that it seldom extended to their own offices and boardrooms.

By contrast, theatre trained designers bring a sense of variation and intuition to their work, which is not shackled by the constraints of precise photometric formulae. This freedom, based on sound technical knowledge, allows a creative crossover from the proscenium stage to the market-place.

Stage and retail have much in common. There is a need to direct attention so that it is focused on what is relevant. For instance, in a food hall, it is crucial that we are aware of the food rather than the walls or floor. In the theatre, we want to be able to see the actor onstage who is talking.

The stage's atmosphere of the dramatic and of the ceremonial is also present in hotels, where there is a feeling that everyone, staff and guests alike is playing a part. A hotel is a controlled theatre of lives and events, contained within a designed environment. Developments in theatre-lighting control have gradually found their way to the commercial sector, and now, in the form of 'scene-setting' and dimming, are available in sophisticated formats.

Figure 3
Atrium of the Banco Santander, Madrid, Spain. A contemporary project by Austrian architect Hans Hollein, where lighting emphasizes the dynamic interaction between roof prism and cylinder. Erco Lighting Limited.

Basic working practices

In addition to illuminating a space, the designer will guide the client through the labyrinth of lighting equipment available on the market. They will advise on appropriate luminaire and lamp. A lighting designer calls a bulb a lamp and a light fitting a luminaire. If the luminaire is visible to the public then this has to be taken into account, as well as the lamp utilized and the light distribution produced. A lighting designer will not specify equipment from catalogue data alone. A luminaire will be tested and examined at first hand. Crucial factors, such as lamp performance and value for money, are taken into consideration, in parallel with whether the design is suitable for the intended interior.

A manufacturer may claim that lamps can be easily replaced by the such-and-such method which they have patented, but fail to explain how in fact this actually works. Fittings can be incorrectly installed by the contractor, and the designer must be available in order to avert potential disaster. Due to financial pressure or geographical distance, this is not always possible. It is not unknown for wall lights to be fitted upside down or with unsuitable lamps. Fixtures can arrive on site damaged or do not arrive on site at all. Manufacturers can promise to deliver on time but then fail to meet the deadline. This can be a real problem for the designer who then has to source an alternative fitting, usually within twenty-four hours. Specifiers depend on the reputation of manufacturers whom they know by experience will be less likely to let them down. It is important that manufacturers publish photometric data and luminaire dimensions, as well as stunning art photos of the fitting.

If you are thinking of specifying a luminaire, then a working sample should be requested. What can appear wonderful in a catalogue sometimes materializes as a piece of badly constructed plastic tack. Catalogues are not always concise about what they are selling. In addition, sometimes luminaires are a style statement rather than a working piece of equipment. When you receive a sample from the manufacturer, take it apart and find out how it works. This will give you an idea of construction quality. A problem spotted will save time later on.

If possible, the designer should be aware of luminaire products throughout the world. If a job comes up in Borneo, they should be able to locate a dozen suppliers in the Far East. Knowledge of a second language is a valuable key to finding new work as there are some countries where English is not widely understood.

The designer should have a technical base to their knowledge, as well as being familiar with ergonomics, modern production methods, materials of construction, and have a grasp of art history and social studies. There should be an appreciation of the nuts-and-bolts reality of construction, and the more subtle laws of design aesthetics. At times the designer will be asked to create a unique luminaire intended for a particular situation which is not covered by the existing light products available on the market. This is called a 'special'. It is then that a knowledge of construction and materials is important.

The artist who is sensitive to the natural world has much in common with the effective lighting designer. Yet even with the advances of light technology, the most skilled manipulation can rarely match the beauty of natural light. Snow, clouds, dust and rain affect the way we perceive a scene. In a theatre, with the aid of scenery and gobos, it is possible to create an approximation of nature. Each designer develops a personal language to help represent this other world. It is an absorption of many elements, taken from knowledge and imagination. The designer should gain as much from a study of Gericault, as from a complex research paper on 'The Nature of Human Performance'. They should be conversant and aware of the latest technology but not overawed by it. The lighting designer is in truth a 'Renaissance Man', who is able to comprehend the language of both artist and artisan.

Lighting designers work with architects and engineers who often use quite separate vocabularies. It can sometimes take several meetings before the various

Figure 4
Tower of the Winds. Toyo Ito. Yokohama. An interactive illumination membrane which responds to changes in the natural environment. Erco Lighting Limited.

factions understand what the others are trying to say. The architect may be more used to working with an electrical engineer who has a particular work approach and attitude to project deadlines. At the start of a project, the lighting designer may be misunderstood by engineer and architect, especially if they are reluctant to question any aspect of the job which is unclear. It is important to explain to both architect and engineer exactly what services you are providing. Method of fee payment should also be understood. It is essential that everyone is clear on how much you are providing and by what date. Consider in advance extreme conditions or emergencies that might develop and prevent you from meeting obligations.

The reason for the appointment of a lighting designer to the team will vary. It is seldom the case that the lighting designer is contracted simply to supply attractive lighting. Both engineers and architects believe they can provide that, and on some occasions do. One reason may be that the architect is running out of time and the project has developed complex constructional difficulties. Perhaps the client requires a sophisticated lighting solution, involving extensive computer-driven lighting control. In some cases, another lighting designer might have been hired but failed to meet cost restrictions. On the other hand, the architect may have worked with a lighting design team before and was delighted with the way illumination enhanced and revealed the building.

The lighting designer has therefore to perform on what appears to be several tightropes at once. Sometimes you will be brought in late on a job, which may mean inheriting a backlog of other people's problems. As the appreciation of the worth of lighting increases, hopefully the tendency to view lighting as a bolt-on element will cease.

Presentation of ideas to the client in a concise and understandable manner is of great importance. With good presentation everyone knows what to expect of you. Keep in regular contact with other members of the design team and tell them what you are doing. Use models and illustrations in addition to clear technical data, so that your concepts are communicated on several levels. In this way the team can understand the design route that you are following.

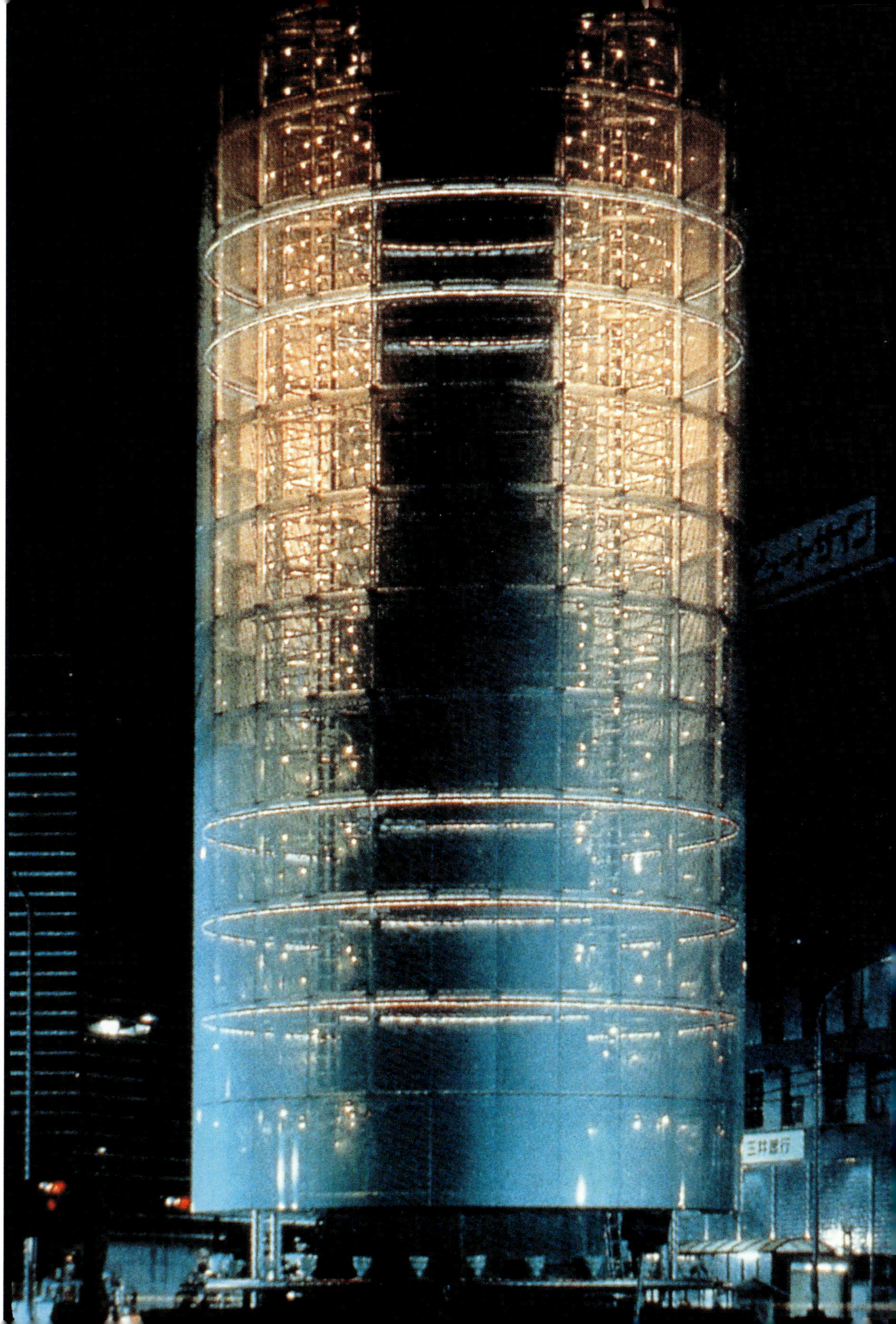

The right lighting in the right place

Figure 5
Tower of the Winds. Toyo Ito.
Yokohama. Erco Lighting Limited.

The influence of light on human behaviour is now widely accepted. Shopping malls use lighting techniques to lead customers from one area to the next. Specialist lamps are chosen to emphasize colouration of meat, fruit and tinned products. Areas where jewellery is displayed are illuminated in a different way from food and clothing.

Certain types of light are associated with particular spaces. You would not expect to find the interior of a church lit with the banks of fluorescent light found in supermarkets. It is in the retail area that effectiveness of lighting can be judged with some accuracy. This manifests in better staff performance, less days lost through illness, savings in energy consumption and increased profits at the check-out desk. It is important that the correct type of lighting is used for each retail area. The first choice to be made is in whether a warm or cool light source will be used.

For example, fridges and freezers would lend themselves to a cool display light, which would emphasize the nature of their use. Diamonds would also look even brighter under a cool accent light, contrasted with a softly illuminated background. Ambient background lighting should be controlled carefully so as not to cancel accent lighting. A furniture display might benefit from a warm lamp source which brought out the natural grain and texture of the wood. Tweed clothing would be made more attractive by being seen under a warm light source. Glassware would be helped by a cool source which would heighten the sparkle of crystal.

Illumination ratios between merchandise and circulation routes can be adjusted to concentrate attention upon the goods displayed. Given a choice of two entrances leading to an area beyond, customers will usually choose the one with a higher level of illumination. If one wall in a room is more brightly lit than others, customers will prefer to sit facing that wall. In a theatre, the stage curtain prior to performance will be more brightly lit than the surrounding walls. This gives people entering the space a immediate signal as to the direction they will be facing. In an unfamiliar space, people prefer to be directed and are often confused when presented with an abundance of choice.

While in the home, we instinctively adjust the light so that a particular activity can take place. If we wish to relax and view television, then distracting lights are turned off. If a special friend is arriving for dinner, then taper candles are lit. Most people like to be in an ordered, relaxed, space. Rooms can be a good reflection of a person's mental and emotional state. Lighting designers have to find the right emotional ambience for each environment. As in art, it is a matter of experience and discernment, rather than the rigid application of hard and fast rules.

Similarly, in a store or mall, if the customer feels an affinity with the space and is therefore happy and relaxed, they are more likely to purchase. In a hotel, if the guest feels at home, yet intrigued by the surroundings, they will be more able to enjoy the experience. They are away from familiar surroundings but still feel safe.

Researchers have for some time studied human response within controlled lighting situations. This data helps designers create appropriate lighting schemes for a wide range of interiors. It must be stressed that often the touch that makes a lighting project exceptional cannot be located amid the relentless graphs of illumination researchers. It is the dexterity of the human imagination that confounds all form of calculation.

Lighting is the most crucial factor within every interior. Light can be used to stimulate or to calm. It is the means by which we judge space. Bland interiors with ineffective lighting produce a negative roomscape which we connect with sadness and failure. Such rooms reflect badly upon the people who exist in them. In such spaces the rich variety of human experience is washed down to a tone of grey.

Lighting designers have to be fluent in many languages both tangible and intangible and, like a stained glass window, be poised between visible and invisible worlds. It is this tightrope between the power of imagination and the fine edge of technology which provides a challenge. It is a balance between aesthetics and mathematical logic absorbing all the creative diversity inherent within these disciplines. Such is the challenge and the journey, though arduous, guarantees adventure.

Figure 6
Time-Space, relaxation, stimulation centre, developed in collaboration with Cologne artist Victor Bonato, management training company H.G. Rupp and interior design office of Schwitzke and Partners. Erco Lighting Limited.

1: Daylight

Such is the abundance of daylight that architects for centuries have attempted to control its access to their buildings. Filtering devices have varied from Arabic carved window screens, to striped Neapolitan canopies and electronic fenestration controls on-line to energy management systems.

We experience interiors through vision, which depends on the way light reveals form and void. Light may emanate through the bars of a prison cell or the frail glass petals of a Gothic rose window. The way in which light is allowed to enter an interior shapes our spatial information. The architect develops ideas regarding function, form and fabric, which relate to the purpose of the structure, the composition of space and materials of construction. Within the tension of these voids, the drama of light is given life, through calculation and vision, enhancing both space and form.

Architecture in daylight

Figure 1.1
Plenary Complex of German Budestag in Bonn. Natural light patterns cast by roof fenestration. Lighting Design, Behnisch & Partner, Stuttgart.

Daylighting and architecture are inexorably linked, though an estrangement occurred with the invention of the fluorescent tube. These brought electric light to industrial workplaces, where daylight was viewed as an unwelcome intrusion. Windows were often opaque so that workers would not be distracted by the outer world. The concept of employee maintenance had not yet dawned upon the captains of industry.

As the windows of Baroque cathedrals influenced the perception of artists such as Caravaggio and Rubens, so the plate glass of the twentieth century reflects the preoccupations of Freud, Hopper, Bacon and Hockney. Architects such as Schindler, Frank Lloyd Wright, Le Corbusier and Eero Saarinen have explored the inherent beauty of natural light. Before the introduction of fluorescents, daylight was a crucial factor in the construction of buildings. Today, at a time when nations are anxious to conserve energy, daylight is once more on the ascendant.

One of the first architects to orchestrate natural light was Gian Lorenzo Bernini (1598–1680) who brought to the seventeenth century a passion for spatial dynamics which would seldom be rivalled. Bernini's fugue-like climactic fascination for forms sculpted by light wove lucid complexity within a controlled architectural continuum. His technique was a massing of surface and void, which relied on sunlight to express the sensation of textural boundaries afloat within vast interiors. This articulation of arches and columns, illuminated by arrays of windows, created rhythmic patterns, constantly changing with the sun's progress across the sky. Bernini was a magnificent sculptor and considered his buildings to be an extension of that art.

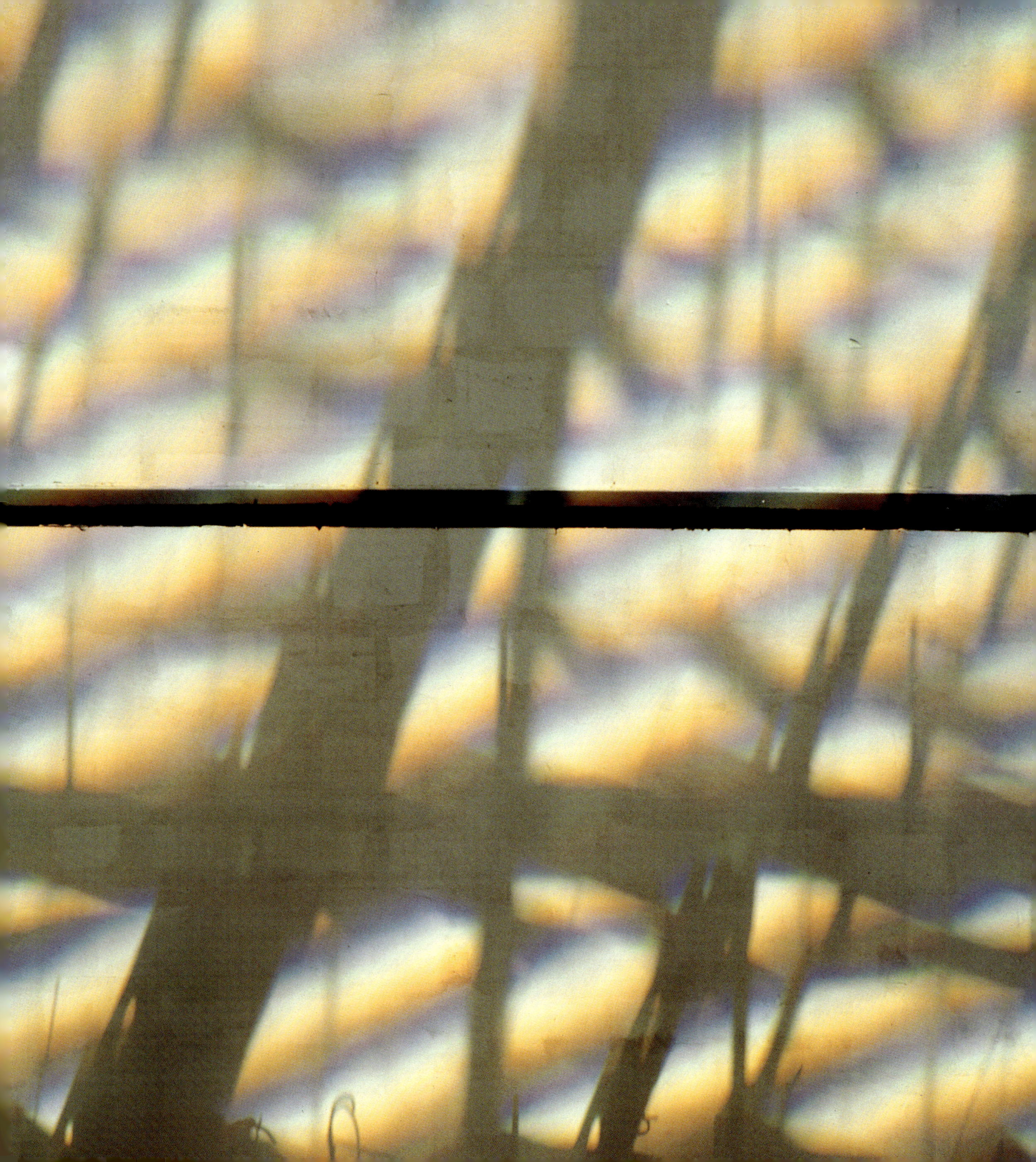

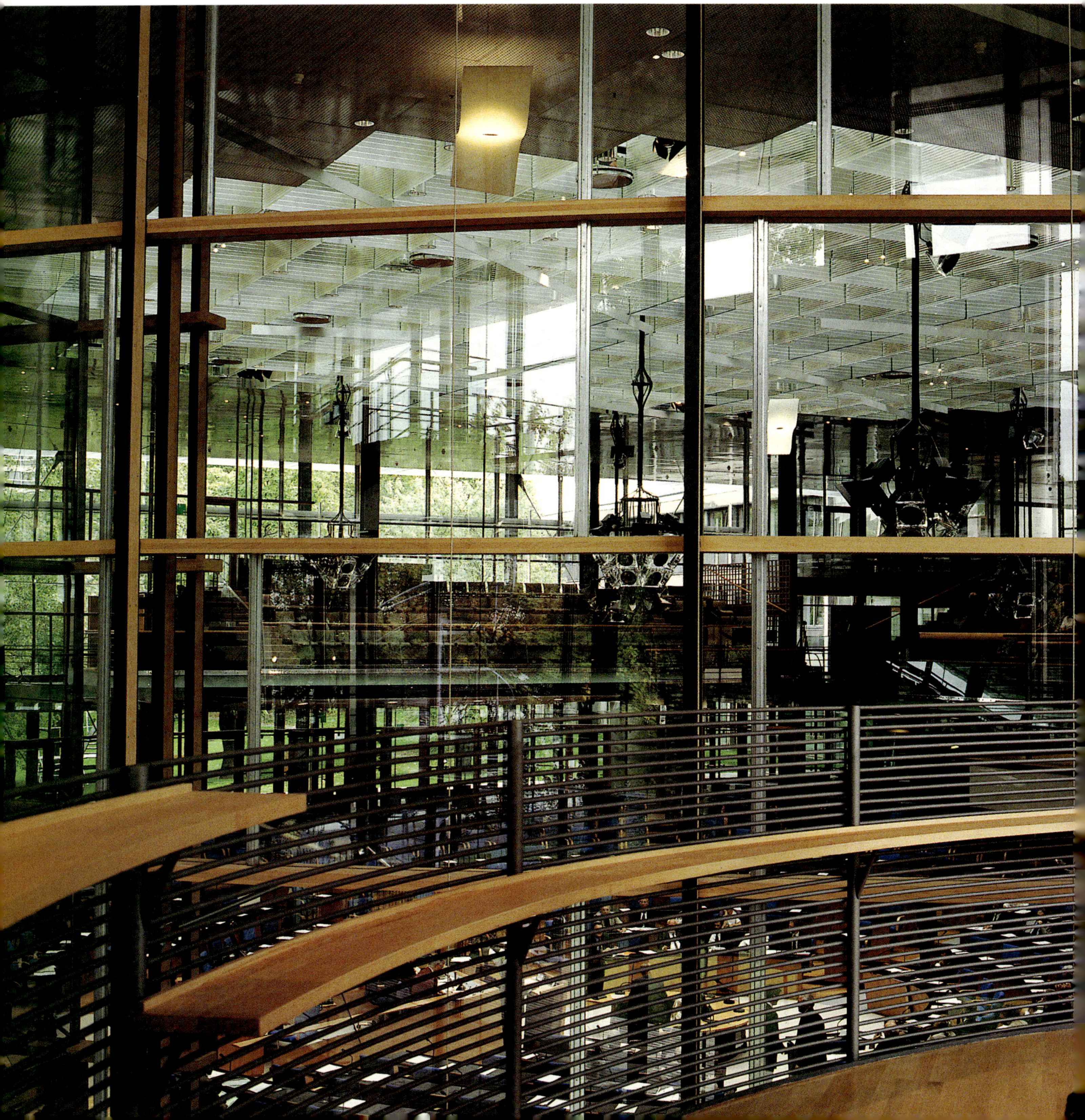

Figure 1.2
Plenary Complex of German Bundestag in Bonn. Careful use of natural light allows expression of interior architectural form. Lighting Design, Behnisch & Partner, Stuttgart.

Bernini integrated architecture and sculpture in a way that explored the fluidity of light. His altar 'The Ecstasy of Saint Teresa', in Santa Maria della Vittoria Rome, conveys a spiritual intensity which is both sensual, yet transformed and remote, with figures of angel and saint, adrift as visions in space. The altar is lit by concealed windows so that a subtle interplay of illumination suggests yet never wholly reveals. It would be interesting to speculate on what Bernini might have achieved with the equipment available today. Today, designed lighting could be described as a compromise between calculation and imagination. It was similar in Bernini's time, though to miscalculate the position of a window at the heart of a great cathedral, cannot really be compared to the incorrect alignment of architectural floods.

The light filtration system of the seventeenth century was stained glass. Many methods have been developed to reduce glare and solar heat gain, while establishing comfortable levels of illumination. Technology races on, but the challenge remains the same, eliminating glare, while optimizing levels of daylight. A number of ways can be used to achieve this. Glazing can be tinted or coated. Curtains are usually only used in a domestic situation, where attention to lighting control is rare.

Exterior screens can include louvres, shutters, or vertical blinds. Screens can be programmed to respond to solar sensors, so that on an overcast day, blades would open fully, and if needed, artificial light would be activated. Atria can be designed as solar-reactive membranes, with temperature variables measured, prior to light-screen angle selection. Traditional fixed devices such as canopies, shields and baffles, have a set response to patterns of sun and cloud. A computerized fenestration control responds instantly to exterior climatic situations. Sensors constantly track the sun. Louvres react to block direct solar rays, while allowing usable reflected light to penetrate the building. At night, louvre shields can be shut down to maintain a continuous thermal envelope, reducing heat loss. Fenestration intelligence is fed to the building's energy management computer, so that a holistic needs profile can be monitored. Of course, the initial cost of such systems is expensive and may only be possible in larger corporate structures.

Daylight in office spaces

Through the centuries we have become used to warm light sources such as fires, candles, gas lamps and domestic bulbs or GLS lamps as they are sometimes known. In terms of visual experience, the coldness of a daylight lamp is a recent phenomena. Introducing daylight to windowless office spaces illustrates the problems that come with cold lighting. Lamps are available which reproduce the daylight spectrum, though they cost more than conventional fluorescents. However, people have complained that these daylight tubes appear chill and austere. People dislike working in spaces without windows. The anxiety of daylight-deprived office workers is a reaction to being starved, not only of daylight, but of a view and sense of space. It is not sufficient to install a daylight spectrum lightsource and expect the problem to be solved. I have seen powerful metal halide uplighters installed in windowless spaces. The employees complained of feeling like laboratory rats. Often there is a 'hot spot' or concentration of light on the ceiling, which indicates inadequate reflector design. The light is shot straight up in the air, devoid of adequate distribution. These lamps give good colour rendition but are prone to flicker, which can be most uncomfortable in a restricted space.

No doubt from a misplaced desire for impact, specifiers often require a bright glow, or white heat, to come from the building's core, usually a metal halide lightsource. Unless properly distributed, such substitutes for daylight will do little for the visual comfort of employees. They are however, excellent, for indoor plants. In combination with daylight, metal halide lamps positioned beneath atria, for instance, support plant life, due to their spectrum of emitted light.

By using metal halide in windowless spaces, specifiers hope to establish a look of efficiency, with added benefits of extended lamp life. Their concern is with what the scheme looks like, rather than the effect it will have on occupants. To blend daylight with lamps of good colour rendition and slightly warmed spectrum brings us closer to the essential human needs of space and sunlight.

It is not necessary to understand the complexities of solar geometry to appreciate the vivacity of sunlight, and the way it transforms a drab space. In a business environment, where quality of light is paramount, a solar light pipe might be the answer. This will reflect daylight down to the centre of a building. At initially greater cost, natural light can be piped to sunless offices. This method has some drawbacks, for example in countries with little direct sunlight, though on dark days full-

Figure 1.3
Plenary Complex of German Bundestag in Bonn. Integration of interior-exterior natural elements creates vibrancy within space. Lighting Design, Behnisch & Partner, Stuttgart.

spectrum electric light can be projected down the tube. Light-transmitting systems use sun-tracking heliostats to collect daylight, which is then reflected down the pipe to windowless areas. In a commercial situation, light pipes would be used in conjunction with electric light. Light-pipe systems are more expensive than installing full-spectrum electric light, but it should be remembered that benefits will last for the life of the building.

So fundamental is our need to be near exterior space that great ingenuity has been exercized in the construction of artificially-lit atria, skylights and stairwell domes. It is important that architectural detail around artificial windows be kept as close to authentic scale as possible. Louvres, lamps and prismatic diffusers should not be visible, for the light must appear to radiate naturally from the window. Fluorescent tubes in the cool spectrum, giving good colour rendition, help suggest the appearance of daylight.

When illuminating a simulated rooflight, it is most important to avoid patchy tonality. Illumination, whether through glass or aperture, must be consistent. It is possible to install false windows in sunless, core, office spaces, and several manufacturers have experimented in this area. It has been found that choice of wall and flooring materials, in conjunction with pleasing light, contributes to a non-claustrophobic environment. It must be considered how much employees should be calmed by lighting and to what extent stimulated.

Daylit atria

Each building will present contradictory difficulties, depending on orientation, glazing, heating systems and insulation. In atria, there is great potential for creative management of daylight. Combined with natural light, indoor planting and landscaping encourage a mood of expansiveness. This can help soften the visual scale of commercial atria. If several species of plants are required on multiple floor levels, then light will have to be sufficient for shrubs and flowers with different illumination needs. Computer modelling techniques can evaluate specific design parameters and provide accurate predictions on how internal chiaroscuro will be patterned.

In terms of scale, atria are an extension of the glass-arched railroad stations. Now the temples of steam have transmuted to the multi-functional malls of modern consumerism. Properly designed atria are both beautiful and functional and fulfil social, economic and physical needs, while providing a reassuring interface with natural elements.

Hotel or office atria can be used as interior circulation routes. They can also provide daylit café or rest areas, which are sheltered from wind and weather. During winter months metal halide lamps provide good colour rendition for architectural surfaces, and excellent irradiation to promote healthy atria plant growth. Different wavelengths of light have different energy properties. The processes of plant growth – photosynthesis, photomorphogenesis, photoperiodism – require particular wavelengths of light. Some lamps will not be able to stimulate the full sequence of growth. Environmental factors such as temperature and humidity, along with exposure to daylight, are of importance when nurturing plants.

Figure 1.4
The new Opera House in Helsinki, Finland. Natural light is supplemented by artificial light, used to accentuate focal points. Erco Lighting Limited.

Maximizing daylight in small buildings

In smaller buildings, where budget will not provide for automatic louvre control, there are several ways daylight can be utilized. Glare-blocking overhangs and light shelves are cost-effective ways of making the best use of available natural light. A light shelf protrudes both inward and outward through the window, above head height so that daylight is reflected evenly into the space. The surface can be coated with reflective paint or in some cases a specular metal coating.

The interior edge of windows can be rounded to reduce contrast, between daylight and interior wall. Windows should extend up the wall as far as possible, though calculations should allow for the negative factor of heat gain. On the ground, exterior reflective surfaces such as paths, patios and pools, can reflect substantial quantities of light into the building. The undersides of projecting light shelves can be white in colour to reflect light back towards the windows. On the upper windows, if there is no shelf to act as direct sun shield, white slatted horizontal blinds can be used to block direct sun, and bounce reflected light to the ceiling. If there are full blinds, they can be shut down at night to maintain an insulation envelope.

Skylights are more reliable than they were, and less prone to problems of leakage. Direct solar rays can be avoided by using large, vertical, north-facing skylights. Improved levels of daylight, especially in winter, will help to ease worker fatigue and seasonal depression.

Glass blocks are another means of allowing daylight into a building, while creating a visual screen between interior and exterior, which can be useful if privacy is required. They are effective for deadening noise, and have good thermal insulation qualities. Glass blocks can act as light shelves, the horizontal surface between blocks reflecting daylight up towards the ceiling. Floor-mounted uprights with beams grazing glass partitions can add attractive sparkle to the uniform tones of the office environment.

Figure 1.5
Carre d'Art, Nimes. Architects Sir Norman Foster and Partners, London. Lighting Design, Claude Engel, Washington D.C. Rooftop louvres diffuse direct solar rays. Erco Lighting Limited.

Fabric structures

For centuries, canvas and fabric structures have protected people from the weather. Fabric provides diffuse light and excludes glare and synthetic fabrics can withstand extremes of heat and cold. Fabric can be stretched and sculpted to create an attractive structure, which can be used for art galleries, sports stadia, atria covers and goods storage. The amount of light entering can be controlled, depending on the transmittance factor of the fabric. Systems can be supported like tents, suspended, inflated or framed on geometric modules. This flexibility enables a large area of ground to be covered in a short space of time though, obviously,. in some climates consideration must be given to rain drainage, wind speed and heavy snow.

Electric light would be needed in fabric-covered exhibition or retail spaces, though overall energy consumption would be less than in a hard-shell building. Cost savings are apparent when fabric structures are used for storage. In such use, problems of nighttime heat loss are not so critical.

Fabric, stretched over hotel or retail mall atria, provides diffuse natural light and weather protection. The future for daylight systems indicates an increase in reactive dynamic glass envelope structures, designed for energy conservation, with flexible personal lighting options, for those working behind the programmed scenarios of automated fenestration. Electrical engineers, and some lighting manufacturers, tend not to be optimistic about the potential of solar-powered systems, dismissing them as expensive to install and prone to breakdown, though this may partly be due to a subliminal fear that electricity, their only lifeblood, might at some future date become redundant.

Daylight and energy conservation

Environmental conservation is now viewed as a necessity, but the use of solar energy is still at a relatively early stage, and the cost of gathering systems will have to decrease before they are widely specified. The advantages of daylight systems and passive solar buildings are considerable, and should not be judged wholly on initial installation cost or payback time.

Architects were the first to appreciate the luminous variations of natural light, allowing it to reveal their buildings in accordance with craft and vision. To only measure light through the calibrated graph of a computerized design tool, is to ignore the emotion, intuition and sense of place which can transform an adequate void to one which is sensational. A room will usually be described in a subjective way. It is light which enables us to interpret space; it should provide more than an illumination of our line of exit, and a means by which to avoid the furniture.

Figure 1.6
Marine Simulator centre, Rotterdam. Architects, Sir Norman Foster and Partners. Cool illumination defines the void between curve and plane. Erco Lighting Limited.

2: Retail

The changing marketplace

The urban shopping centre and mall share much in common with the earliest markets and covered arcades. Glass arcades owe their origins to Greek and Roman vaulted passages which housed purveyors of fish, meat, metalwork and other essential produce. It was not until the end of the eighteenth century that shops replaced wooden shutters with glass windows. Painted trade symbols gave way to arranged windows which catered for the tastes of the emerging middle class. The Industrial Revolution brought a new range of merchandise and customers with money to pay. Competition for this business produced blatant advertising, as merchants jostled to sell the most dubious of inventions. Light was provided by gas lamps and candles.

Window-shopping was enjoyed as a form of entertainment. Horse-drawn trams, mud and yelping dogs did not distract the potential customer who patronized the glazed arcades. These covered walkways became not only a showcase for merchandise, but a display ground for consumers.

In the twentieth century, the invention of refrigerators and automobiles enabled the development of American supermarkets and their imposition of out-of-town centres, accessed by a newly built system of highways. In time, this US model was replicated on a lesser scale throughout Europe. Today, malls serve the multiple needs of all levels of society who have access to transport.

As supermarket changed to hypermarket and then to mall, it became economically desirable to maximize daylight and reduce escalating electricity costs, thus designing increasingly larger glass envelope structures. Malls were not subject to the same crime and vandalism as city centres, in part due to their distance from areas of inner-city deprivation. Also, the extensive use of glass made their interiors highly visible. At night they were brightly lit and often patrolled by private security police.

Due to the vast scale of some malls, many people experience a sense of alienation; sympathetic lighting of inner circulation voids can help counteract sensations of spatial exposure. Malls are a far cry from the gas-lit arcades of the Victorian era, yet it is such enclosures of intimate warmth which many architects seek to conjure within the retail consumer satellites of the urban outscape. Some older city-centre buildings have been renovated or internally rebuilt, to replicate in part the gas-lit glazed arcades, whose mix of romance and history reassures the harassed shopper.

The flickering fluorescents which marred the first generation of supermarkets are now reserved for the back-street purveyors of auto parts. Fluorescents were cheap, and their wall-to-wall 'fitted-carpet' lighting suited market owners whose only concern was that the customer should be able to see the merchandise. Today the consumer youth of Europe and North America is media-literate, raised on television advertising, with a low boredom threshold and a need to be visually stimulated. In the malls of consumer retail, light is a medium sufficiently broad to encompass both grandiose gesture and intimate counterpoint.

Illuminating the product

All retail lighting is an attempt to lead the customer to the product. This is as true within a small shop display as it is in a multi-outlet atrium. Watches gleam beneath a single spot, or a canvas canopy glows from within, charting customers across a crowded mall. A shop lacking conciseness of mood is a somewhat dull and scentless flower. I am not speaking of tracked spots, aggressively angled, but of the quality of the illuminated surface, an emphasis on the light, rather than the means of projection. A perfume counter, for instance, can reflect a palette of austere charm, delicate light glazes framing packaging, while accent beams reveal the product name. In such a situation, low-voltage tungsten halogen adjustable downlights might be chosen for their compact size, wide selection of beam angles and good colour rendition. In a more restricted display space, fibre optics might be used for highlights, with dimmed tungsten halogen downlights to create a soft ambient glow.

Figure 2.1
'Bagstore' shop in Dusseldorf, Germany. Subtle use of shadowing creates warmth and pattern within open floor plan. Lighting design: Professor Wolfgang Doring, Architect.

In narrow windows or display cases, fibre optics can provide a cool-running, unobtrusive, versatile light. They cannot as yet generate powerful illumination, but this is a limitation which research may overcome. They can be used as starpoints in suspended roof structures, to help counteract the 'black-hole' effect of unlit atria at night. This creates a sense of enclosure around the mall.

Within display units, the light points of fibre optics illuminate the product while the lamp source is remotely located some distance away in a convenient floor or wall space. Fibre optics can be fitted to the underside of handrails, used as signage or downlights. The actual light point radiates no heat, making them ideal for exhibitions where fragile items are displayed. Fibre optics can bend around corners, and in the side-light emitting version can draw around architectural forms, in a manner similar to neon. Used in conjunction with other light sources of greater power, they provide a versatile lighting tool.

Goods on display must be protected from the heat and light given off by many lamps. Spotlights, downlights and wallwashers can be fitted with ultraviolet filters, louvres or glass protection shields. These are advisable in most retail situations. If a lamp explodes and glass is scattered, there may be no physical damage, but the psychological impact will be negative. Radiation from lamps can cause fabrics to fade. Furniture restorers sometimes use metal halide lamps to prematurely age fabrics, so that they match furniture. Low-voltage or metal halide lamps within closed display areas can generate excessive heat. Display spaces should have adequate ventilation and care should be taken that lamps are not placed too close to display merchandise, or sales staff for that matter.

In city centres, window display will be governed by street illumination levels. The lighting approach of properties on either side of the outlet should also be considered. If there are stores adjacent, with similar levels of illumination, a no-win situation may arise, where one window cancels out the effect of the other. Lamps of good colour rendition are required so that passersby can appreciate both colour and texture. The public like to understand what they see. This does not mean that lighting must be bound by orthodoxy. However, it is desirable that a shop's ethos or character be clearly communicated to the public.

Merchandise should be considered in relation to the display backdrop. A dark green background will affect the perception of a burgundy shirt compared with primrose or warm grey. The display lamps used will introduce another variable. You may have bought a sweater only to find on returning home that the colour was not quite what you remembered. It will not appear as vibrant under the light of a domestic bulb, which has a dominant yellow colouration as opposed to the crisper white light of metal halide or tungsten halogen display luminaires.

Willkommen im bagstore

Illuminating the emporium

If shops are expressions of what the public wants most, then they are indeed mirrors of our lives and desires. In times of depression they are gloomy, while in times of growth they exhibit a diversity of styles, which reflect the needs of both old and young. In most city stores, there will be a lighting 'set up' for use in daytime as well as night, while in a small-town shop, the single row of spotlights may be switched off at dusk.

In cities, the retail commercial hub can alter according to demands of the public and whims of the developer. In a short space of time, rents can become extravagant and out of touch with business reality. Traffic flow is changed. A national outlet relocates. What may one year be a viable business can be bypassed by the unpredictable shifts in market forces. Stores can, like the arcades of De Chirico's paintings, be made ominous by their sudden emptiness. While good lighting cannot be expected to circumvent such twists of fate, it can take a moderate business and turn it around to a successful one.

Lighting designers, eclectic composites as much of their work reveals them to be, must not only gauge the appropriate lighting, but take into account the economic and geographical profile of the premises. A shop selling cut-price sportswear will, for instance, require an approach which reinforces a no-nonsense streetwise image. A bright but cheerful overall scheme with select spotlighting will probably be all that the budget will allow. The need for gas-powered, iron-scrolled, flambeaux will be unlikely. On the other hand, the florid sentimentality of a retro1950s ice-cream parlour may demand a peppy chrome ambience which draws attention to the decorative light fittings, in contrast with today's emphasis, which eschews the concept of luminaire as star. An exclusive shoe boutique may require the warmth of tungsten halogen in alliance with an opulently orchestrated contrast of cool metal halide.

From an academic or didactic standpoint, the aesthetics of lighting appear to exist on a constant edge of seeming contradiction; it is perhaps this inherent tension which weighs certainties of science against the visual conjurations of craft, but always with a hint of rules that can be made to bend. There are rules but some of them cannot be captured on computerized lighting scenarios. This is true, for instance, when balancing warm and cool light sources within the same interior. A good eye is more essential than a pile of photometric readouts. Interactive calculations are of course invaluable, and they do have their place, but the human eye is still more sophisticated than the most complex electronic visualization tool. There are of course persons unblessed with a 'good eye' who believe that all lighting problems can be solved by data.

Today the reality of shopping can lead to a battering of perception. It is often not a subtle blend of products which surround one, but a beast with many heads, each one attempting to howl down the others. The active ferment of retail requires that the customer be piloted among the shelves without sensations of merchandising meltdown. The idea is to promote sales, but customers can get confused and unable to find the goods they wish to purchase. In extreme cases they become unable to even find the exits. Lighting must provide guidance, stimulation and a positive sense of direction. In European retail, there is a tendency to illuminate brightly at key pivotal points, such as the ends of display shelves or the tops of escalators. This method is a marker-buoy system which connects channels of even, ambient, light with intensified pivots of accent light. Levels of brightness between accent and ambient light will vary with designer and the national preference. Ambient light provides an even background or tone, a visual primer upon which are layered modelling and highlight.

In stores where overall illumination is what budget allows, light levels should enable the elderly to see the labels on packaging and luminaires should have efficient reflectors and louvres for even light distribution. Carefully graded intensities of lighting can be zoned to create some visual variety in what would otherwise be a uniform light space. With more money available, a linear suspended system can provide both ambient and accent light while lowering sightlines in a high or ungainly structure. Suspended systems are also used when retaining a ceiling which is of historic or architectural interest.

Accent lighting

Accent lighting should attract the customer without having to wash out the product with a row of spots. It is possible to overlight a display so that it becomes difficult to look at. In situations such as malls the walkways will generally have a lower level of illumination than the front of the retail units. Suspended ceilings are the most common lighting support system used in retail as they will accept a wide range of luminaires; tasks such as lamp-changing can be accomplished with little difficulty. Accent and ambient luminaires can be changed around on a suspended ceiling if product needs to be rearranged on the shop floor. The colour temperature of an accent luminaire will be chosen to enhance the object lit. The market image of the product will affect the degree of modelling required.

An electric drill can be lit with the same attention to surface and form as a Murano vase, but in practice seldom is. To what extent is a sense of surface texture crucial to our appreciation of the goods displayed? Should the item be displayed dramatically, perhaps in a stark, back-lit, silhouette, so that attention is focused on the shape at the expense of surface detail?

Usually a compromise is reached between shape and surface with texture or lettering visible but sufficient form displayed to suggest three dimensions. As in portrait painting, the relationship between dark and mid-tones is animated by the dextrous placing of highlights. And, again as in painting, the overuse of accent luminaires can wash out the three-dimensional depth we seek to create. Too much sparkle can become unacceptable glare.

Some merchandisers have spotlighting everywhere, so that the ceiling begins to resemble a score of jumbo jets coming in to land. To be able to design spatially, to create a fluid visual field, requires a measured overview. The broader canvas demands that each lighting element should be considered in relation to its significance within a connecting sequence of visual stimuli.

Good modelling and colour rendition is also required for sales counters, so that staff are free of unfortunate facial colouration or shadow. Not all lamps emphasize healthy skin colour. It is important that sales staff look well. In businesses where there is little in the way of visible product, such as health clubs, auto repairs, insurance sales or travel agents, the need for good facial modelling and skin colouration is essential. In a sense, people have to be sold as much as product, though this is a factor which is often overlooked by interior designers. It is unlikely that you would make a high-cost purchase from someone you did not like the look of.

Good lighting is important for the customer also. Hair-styling salons often pay scant attention to their client's appearance in the mirror. Light is directed on the top of the customer's head, for benefit of the stylist, giving the client facial shadows. It is not every day that a person sits for up to an hour staring at themselves in a full-length mirror, and salon owners owe it to the customer not to imprint their memory with an image resembling Frankenstein.

The amount of accent light required to make a product stand out will vary from item to item. Adjustable luminaires which can be dimmed are useful in meeting these visual variations, which are caused by the reflectance value of the object, whether the surface be matt or gloss, and the degree of colour saturation. A brown tweed will reflect as little as 25 per cent of the light trained on it, while a chrome toaster might reflect around 80 per cent. In the display of refrigerator or white kitchen goods, highlights would be carefully applied to mute the glare. Perception is directed not only by illumination of the merchandise but the degree of contrast it exhibits in relation to its background. It would be unusual to find white goods displayed against a bright vermilion backdrop. A warm neutral background would provide subtle contrast and relate to any tonal modelling created by accent and ambient light. In most cases it is advisable that matt surfaces be used in display backdrops. This helps absorb glare and reflection.

Annual changes in style and fashion affect most sections of retail. Though a store may be modern, extremes of luminaire design are best avoided except in unusual situations. A flexible lighting solution should outlive several image changes and the occasional repaint. Most luminaires, unless they are design fads or concept extremes, do not date quickly, though extensive retail use of spotlights seems to have abated. The low-voltage spotlight has disadvantages in respect of heat output, though this is compensated by compact size and good colour rendition. With the technological advances in compact fluorescents, providing lamps with reasonable colour rendition, low heat output and long life, it seems they may eventually provide an alternative to low-voltage tungsten halogen.

The Mall experience

City-centre stores are designed in a tight, product-driven manner, more intense than their near country cousins, the malls. In the latter, a breathing space is allowed, in the shape of atria with plant life and an abundance of natural light, which the constrictions of a city-centre site do not always allow. Malls provide a social focus and sense of spectacle that goes back to the piazzas of Rome. The first malls emulated glazed arcades, being parallel shop fronts connected by a span of glass. The building is often decorated with architectural motifs adapted with almost Disneyesque glee from Greek, Roman and Egyptian styles. In such an interior, luminaires which are design extremes may be exactly what is needed.

At the tops of escalators, a glowing area of illumination should encourage shoppers to ascend. Customers will avoid climbing stairs if at all possible. This aversion to stair climbing has never been fully explained, and was even remarked upon by builders of the Victorian arcades. The fact that escalators move does not seem to make all that much difference. People have therefore to be encouraged upwards. Sparkle points of low voltage or fibre optics, or tungsten lamp globes beside flowering shrubs are all enticements to explore the upper levels. The spatial sequence of vertical movement can be encouraged by locating intimately lit cafés or neon-edged fast-food stalls on upper levels, so that they are clearly visible from the ground. Shoppers enjoy gazing down upon the throng while sipping a pleasant cappuccino. Even on a regular basis some people feel uneasy being far above the ground. Diffuse, warm lighting with perhaps a tungsten halogen source, can help establish a light that is both reassuring and calming. When you think of the scale difference between the usual domestic environment and the average shopping centre then it is not suprising that people, especially old people, can become frightened and confused by an excess of spatial choices, both vertical and horizontal.

Using a warmer light source which hints at sunlight helps create an ambience which is both natural and non-aggressive. Although tungsten is a relatively power-hungry lamp compared with fluorescent or halide sources, this aspect should be considered but not override the undoubted qualities it does possess. Lighting in retail is a major sales incentive and an increase in sales can far outweigh the cost of comfort-effective lighting. It makes sense to invest initially in order to stimulate long-term return.

This highwire act between expectations of customers and store performance depends on matching what is promised with what is provided. Lighting plays a leading role in developing character within the merchandising plot. If a shop sells economy lines that does not mean the lighting should be bad. A shop at whatever level in the income pyramid wishes to sell to as many customers as possible. Good lighting in any situation makes intelligent and creative use of resources available within a given budget.

Lighting systems: maintaining visual coherence

We are still a long way from experiencing the perfect lamp. In truth, manufacturers have at times made optimistic claims for lamps which were not entirely fulfilled. The general public are most reluctant to experiment with new lamp forms and the retail sector receives the cutting edge of lamp manufacturer's publicity. Fluorescent lamps which have been designated white or cool display distinct colour differences when made by different makers. If a shop relamps with a collection of sources named white fluorescent from different companies, then the result can be like a rather badly emulsioned wall.

It makes sense to have a relamp system in which one person is responsible for maintaining a visual coherence as regards the purchase of replacement lamps. Often after an outlet has made several rounds of lamp replacement, and various people have readjusted the lights, management notice that things are not quite as they remembered them. Lighting is the most significant element in retail design, yet management seldom have a efficient lamp-replacement strategy. Would they let just anyone service their BMWs?

Updating and relamping a lighting system requires inbuilt design flexibility, whether in suspended ceiling, track or linear configuration, so that light patterns are adjustable and new-generation luminaires can if required be assigned to the existing grid. A heterogeneous light palette demands that lamp output be balanced to provide optimum photometric, economic and visual impact. It should be possible to provide this update facility without having to replace the entire lighting system.

It is difficult to predict what direction retail lighting will take in the next twenty years. I suspect that the fibre optics input will increase as technology improves, creating more powerful light sources at affordable cost. The active use of daylight will spread as media environmental pressures influence corporate thinking. There are complex economic and conservation issues when projecting retail aesthetics and the 'need to sell' on to the sensitive eco-politics of world energy. There is also a case for sanity conservation, and compromised illumination in office or retail is a major cause of stress and absenteeism. The act of shopping provides a worthwhile social function and people will continue to look for something to buy. The challenge for the designer is to make finding it an enjoyable experience.

3: Hotels

First impressions

It is the new 'designer' and 'retro-moderne' hotels which offer the most exciting and diverse lighting opportunities in both Europe and North America. This is due to the range of facilities provided, which require the creation of several quite distinct interlinked environments. A sense of theatre allows guests to move effortlessly from one prepared 'scene' to the next. Such buildings often have celebrated or historic façades which help prime the guest for the sybaritic world that lies beyond.

In the medium-range business or holiday hotel, this hint of delights to be savoured will not come from the façade, but from the presentation of the reception foyer. Many good hotels have bland exteriors but emphasize quality by a careful attention to detail in the reception area where first impressions are formed. This establishment of identity reassures the guest as to the level of service provided. Lighting, both natural and artificial, plays a key role in accenting the hotel's character. First impressions are formed at the reception. In daytime, light levels in the entrance should be calculated to allow visual adaptation after sunlight. At night, canopy or façade illumination should not dazzle guests moving out towards darkness.

One of the purposes of a hotel is to provide pleasure, and this is difficult when guests are not sure of directions. Accented illumination of the reception desk allows guests to locate it without confusion. There should be clear indication in which direction swimming pools, coffee shops and restaurants are located.

The busy company salesperson does not expect a lobby hung with chandeliers. Neither will they expect rows of fluorescents with ill-fitting prismatic filters. The lighting in a medium-cost establishment should be cheerful and sufficiently varied to suggest a habitable space, rather than a baggage-handling hall.

Adjustable, recessed, downlights and wallwashers are one solution: they are unobtrusive yet capable of providing varied levels and angles of light. When highlighting pivotal points – such as the foot of stairs, reception, entrances to corridors – coloured prints and flower vases can help to lift an otherwise bland overall scheme.

Figure 3.1
Hotel swimming pool. A sculpted minimalist illumination which avoids clinical austerity. Professor Wolfgang Doring, Architect.

Other factors which contribute to a sense of well-being are the use of lamps providing good colour rendition, not only for furnishings but for the guest's appearance as well. If possible, these lamps should be sourced from one manufacturer. This counteracts the variation of light colour which can come from lamps with different origins. In the reception area, if luminaires can be dimmed, this helps provide a subtle range of lighting possibilities. In corridors, the illumination of large prints can help reduce an appearance of excessive perspective. Carefully lit surfaces can add quality to a restrained, budget-conscious, scheme. However, the lighting designer should be consulted at an early stage in the planning process, and not brought in at the last minute as a kiss of life to a botched interior. Miracles can happen but they require preparation. When there are cost restrictions, then of course the answer is compromise. This can only be successful if there is a degree of flexibility between all parties concerned.

Nostalgia and modern lighting systems

The traditional hotel mimics the chateau or country house to a point of almost ecclesiastical gravitas. In this haven, the not-so-aristocratic can ascend towards dreamed-of realms. To this end, the centuries are plundered for heraldic windows and historic tapestries. The industrialist or business executive generally prefers oak to stainless steel and wishes to be reassured by the decorative motifs of a vanished age. Not for them the relentless geometry of the twentieth century or the mean detail of modernism. Quality is associated with mature materials and discernible craftmanship. Such hotels provide an oblique sense of history; perhaps their promise of stability and security is lodged in race memory, and a time when fortified monasteries offered shelter to the exhausted traveller.

Some hotels that had been modernized so many times that their original interiors were unrecognizible, have been subject to extensive reinvention, so that they once more resemble their first grand gesture of existence. It would appear that market forces have dictated that this is what many hotel customers want.

This trend towards nostalgic opulence might appear to create problems for an innovative lighting designer, given the electrical advances achieved since the gas lamp, but just as such buildings are heated by the latest equipment, so too period lighting can incorporate a well-concealed array of advanced technology. Modern lamps can be fitted within historic or replicated fixtures, and connected to microprocessor controls, thus satisfying the retro-aesthetics of period-style aficionados. Though an interior may recreate the monumental hyperbole of the 1890s, there is no reason why the lighting technique employed should not be a compendium of electrical sophistication. Dimming and subtle sequential 'scene changing' can be used to good effect in the traditional interior.

Due to initial high cost, extensive processor-driven illumination is usually reserved for up-market hotels. Using microprocessors, a selection of 'lighting scenes' can be programmed to rotate through a 24-hour sequence, creating a sympathetic ambient atmosphere which responds to the needs and activities of both staff and guests. A core of unifying harmony can be maintained, while some areas such as conference rooms, shops and communications suites would require a manual override facility. The quality of light could also be adjusted, in response to the seasonal changes.

Figure 3.2
Hotel Rolandsberg, Dusseldorf, Germany. A hint of sparkle brings the unexpected to this neo-classical bar interior. Professor Wolfgang Doring, Architect.

Figure 3.3
Grand Hotel, Brighton. Spectacular ceiling illumination creates a distinctive ethereal ambience. Lighting design by Maurice Brill Lighting Design Limited.

Presence-detection light switching is suitable for service and maintenance areas, where there is limited through-traffic. It can also be used in public area toilets during hours of daylight. Guests like to feel in control of their surroundings and a sense of security is essential in stairways and corridors. Detection units which leave circulation areas of the building in darkness are not suitable for the hotel environment. A relevant application is bedroom-occupancy detection. This ensures that lights are not left on when guests have gone out for the evening. There can be substantial energy savings by the use of systems which react to guests' movements.

'Intelligent' luminaires which respond not only to exterior levels of daylight, but to the operational pattern of other luminaires to which they are linked, can help conserve energy in reception areas and circulation spaces with windows. Interactive 'intelligent' lighting saves energy in areas which receive daylight in undependable, variable, quantities. This will vary somewhat, according to the amount of sunshine that can be relied upon. Northern France will require quite different programming from the southern regions of California. The gambling hotels of Las Vegas are one exception in that they do not make full use of available natural light.

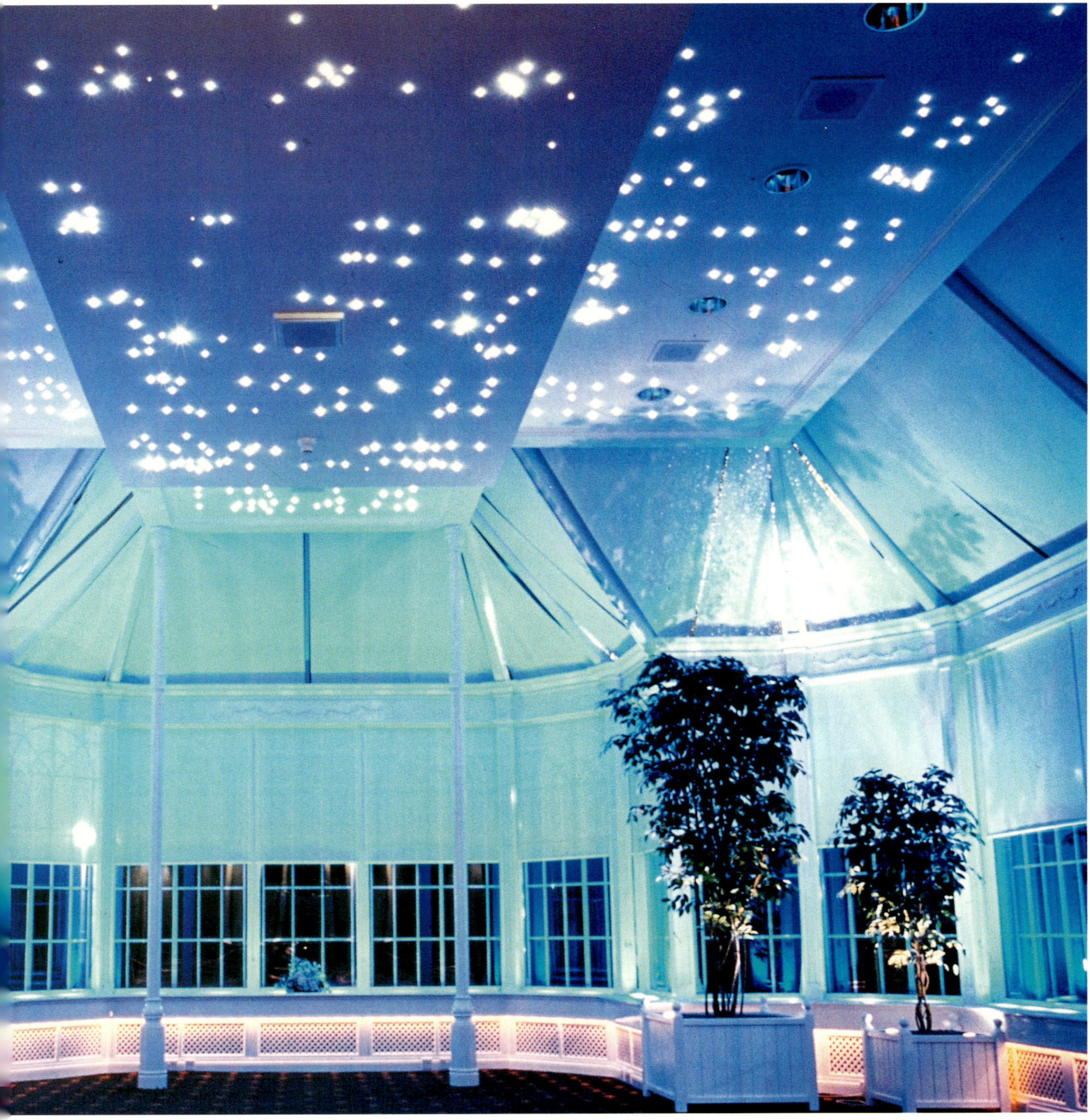

'Warm' and 'cool' illumination

One way of exploiting the dynamic virtuosity of space is by the dramatic emphasis of contrasting warm and cool areas of lighting. Complex vistas can be animated by skilfully juxtaposed phrases of 'warm' or 'cool' illumination. Selection of the appropriate lamp source will depend on a number of factors. These include required lamp life, lamp orientation, lumen output and location within the structure. Light can be washed, reflected or directional, but must be carefully composed so that passages of illumination work both with and against each other, thus creating a tension which releases the volumes of space. In this way intersecting patterns of 'warm' and 'cool' light can be organized to compose a dynamic whole. This type of spatial interplay can be seen in the paintings of Caravaggio, Rubens and Rembrandt.

A 'white' light source, such as metal halide illuminating an entrance canopy, can create a focus of coolness in contrast to 'warm' façade and interior reception lighting. A 'cool' light source can be effective for ponds, waterfalls and fountains and is also applicable for glass and metal spiral staircases. 'White' compact fluorescent lighting can help provide the correct ambience in bathrooms, where the interior decor often includes chrome with black and white tiles. It should be remembered that if the source is too 'cool' it may affect our perception of skin tones.

To give guests an appearance of health is important, and it can be achieved without the use of bronze-tinted wardrobe mirrors! Care should be taken that there is adequate light in the shower compartment. There should be an even illumination of bathroom mirrors so that no facial shadowing is present. The beneficial effect of good bathroom lighting should not be underestimated. It can give a positive image to guests first thing in the morning and this can govern the way they feel for the rest of the day. Bathrooms should not be regarded as units of sterile functionalism: where possible, a dash of colour and optimism should be allowed to prevail.

The partnership of water and electricity has always been an uneasy one. In terms of lighting, there are restrictions on the types of luminaires allowed in a space where baths are run and showers are taken. On a tight budget, where fittings cannot be specially made, this can restrict the design options available. In the more techno-style hotel, the use of fibre optics could be considered. They provide light in a wide variety of configurations but without the presence of electricity. The fibre tails direct light from a remotely located lamp housing. The use of a colour wheel could be an intriguing possibility.

The use of 'cool' light, which indulges our sense of the marvellous, can be found in crystal chandeliers, which irradiate luminosity, transforming the darker and loftier banquet and reception spaces of the traditional grand hotel. Chandeliers should not be dismissed as a period cliché, for they provide a sparkling effervescent light which cannot be achieved by other means. They invoke a sense of focus, brilliance and at the same time luminous dispersion, as if a clear fountain were frozen within space. In large rooms with high ceilings these qualities help overcome problems of scale, and the static blandness of overall fixed luminaire positioning.

Other situations where 'cool' light can contribute are in reception-area surrogate skylight domes and high-level corridor windows, where fluorescents replicate the effect of daylight. In the more avant-garde hotel, you may find a bar constructed from wire-brushed steel and industrial glass. These materials respond to a cooler light source, as will mirrors, drinking glasses and bottles. In this way the bar exists as an arctic vignette enclosed by the conversational warmth of the lounge. Metals such as copper and brass may require a warm light source to fully realize their qualities.

Figure 3.4
Grand Hotel, Brighton, England. A sense of intimacy is suggested within what could have been a characterless corporate space. DeVere Hotels. Lighting design by Maurice Brill Lighting Design Limited.

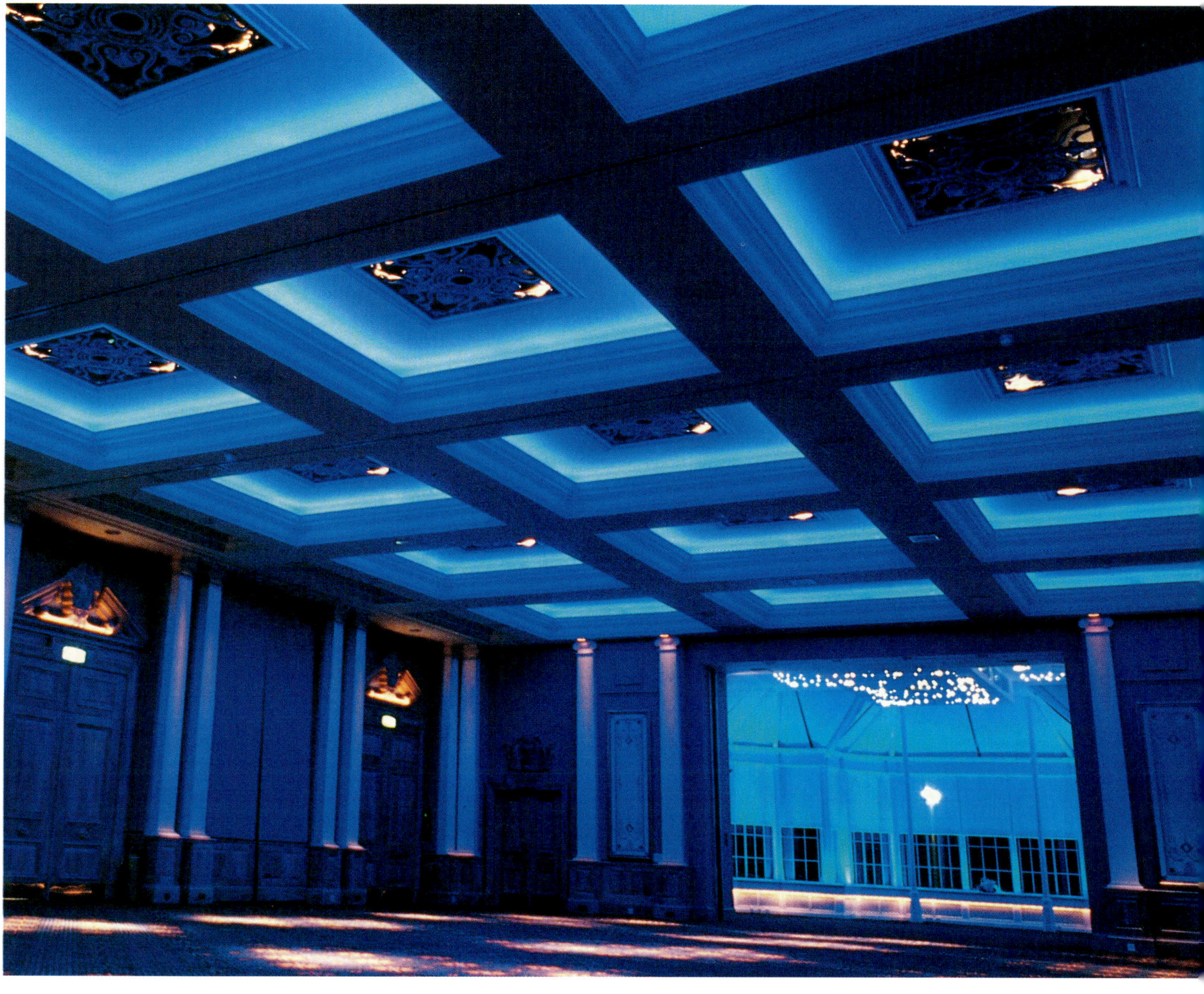

Las Vegas: the oasis

A hotel is not only a cultural crossroads which serves all manner of people and nationalities, but an oasis where commercial pressures can be kept at bay, if need be. Within this complex internal syntax, light brings variation and a sense of spatial resolution which can be both subtle and succint. In terms of the idea of a hotel as oasis or mirage, the 'casino hotels' of Las Vegas are an extreme of hedonistic isolation. Here, mega-dollars are pumped into a glitzy neon artery that leads to a fantasy Xanadu. These hotels' illumination, both internal and external, owes much to the skills of Hollywood's cinematic special effects. The extravagant aerial acrobatics of giant neon signage has given way to computerized digital screens. This is adrenalin-powered speed-read media surfing, where any image is possible, in any combination and at any velocity, so long as it is contained within the screen perimeter.

By day, the city has a decayed reptilian anatomy, as if its spirit had died and returned to the desert. At night, everything changes as the signage begins to circulate, creating a vivid hypnotic skyscape. Such hotels have little use for daylight. Inside, daylight is the simulated afterglow of a cartoon wonderland. Gambling salons have no windows and there are no clocks on the walls as gamblers must not be distracted by the rotation of the outer world.

Adult fantasy hotels do not encourage guests to leave the building. Illumination has to reinforce illusion. This is the type of hotel that pundits in the 1970s claimed would not last. Indeed, these claims are still being made. The hotels and casinos embody a Babylonian narrative style of surreal money magnets where the seduction of gambling is partnered with a military level of surveillance. Cameras operate of such high resolution that they can effortlessly read the seconds display of a gambler's wrist-watch. If fantasy implies a kind of imaginative freedom or escape to Neverland, then it comes at a price to the pocket and personal privacy.

The days of curvaceous desert neon are slowly coming to an end. Digitized signage indicates an interiorization of visual sensibility and a money path to pixel perfection, as the exterior world is increasingly exchanged for a complete absorption in the mindscapes of virtual culture.

The Vegas dream of hotel as 'Wonderland' is largely driven by a sense of place and fate, of actually being in Las Vegas and part of that historic edge of expectation. In the commercial heat of high room-occupancy and jaw-cracking promotion, the need to amaze and the money to pay for it produces lighting scenarios that would not happen in any other hotel situation. It is interesting that in the 1920s and 1930s when motorists' overnight cabin parks were being ousted by neon-crowned motels, there was a craze for theme and fantasy roadside services.

Crude but effective devices, built to catch the attention of passing motorists, took the form of giant hens, toads, dogs, fish and wigwams. Here people could buy ice cream, popcorn and camera film, or sit down for a quick snack. In a sense these popular icons have moved indoors, accompanied by a host of cartoon creatures; like escaped fragments of a child's imagination, these icons have now been reinvented as space travellers and good guy pirates. We have moved into a conjurer's castle of speaking holograms and pixel pyrotechnics. It is amazing how quickly the human mind adapts to change when a roadside snack stop called the 'Big Green Toad' no longer has us diving for the brakes.

Figure 3.5
Grand Harbour, Southampton, England. Lighting accentuates the warmth and surface texture of the floor and furniture while establishing a varied spatial experience beneath the angled void. De Vere Hotels. Lighting Design by Maurice Brill Lighting Design Limited.

Lighting designs for specific hotel spaces

Façade and grounds lighting

Hotel façades should be illuminated at night, not only as a means of promotion but for reasons of safety and and security. In many countries, laws regarding light pollution of the night sky are enforced. It is impossible to eliminate light pollution in urban areas but a great deal of it can be reduced. The spread of sky glow to countryside has meant that it is not just astronomers who are affected, but thousands of rural dwellers who are denied the beauty of the stars. Another factor which should be considered is the waste of energy caused by so much light which is not reaching or is missing the surface required to be illuminated.

As far as hotels are concerned, if pole-mounted globe lights are specified for car parks or entrance drives, they should have the correct optical control, so that light is only directed downwards. Façade lighting should be approached in a restrained manner, grazing select surfaces at a shallow angle, so that a minimum of stray illumination is present, yet form and texture are revealed. Care should also be taken that there is no light penetration of bedroom windows.

As far as light pollution is concerned, hotels are not a main offender: sports stadia, carparks, industrial works and incorrectly distributed street lighting are the principal culprits. There is also a problem with unshielded domestic security spotlights, and this may also affect hotels which use spotlights for the rear of premises. It is more effective to use a low level of shielded light, which does not create excessive contrast, and therefore peripheral blindspots.

Bedroom lighting

The top-class hotel will have varying degrees of bedroom luxury. These must be lit, not only according to their level in the cost hierarchy but in a way that maintains a coherent and effective overall scheme. The transition from penthouse to lobby must not be so marked as to suggest one has descended to a different planet. In the medium-budget hotel, this transition will not be a consideration as bedrooms will be lit to a uniform standard. The business sector that the hotel hopes to target will help shape the lighting criteria. It is important to know who the customer will be and the levels of service they will expect.

Generally, warm light sources are used for hotel bedrooms, whether it be a bare-knuckle roadhouse or one of the sleeping giants of Viennese Secessionism. In the early days of the American Streamline Moderne motel, the most in extras that a room provided was a coin-operated radio and perhaps a bakelite telephone.

Now, hotel bedrooms are multi-functional. There will (hopefully) be a television, video player, radio, and mini-bar and also light switching options for a variety of tasks. The bed should be provided with adjustable reading lights and it should be possible to extinguish all lights from that position. Guests should not require training in switching control as all operational functions should be clearly indicated. Some of the more design-driven hotels may use fibre optics as flexible bedhead reading lights. Others have fixed wall-mounted diffuse luminaires and, in the budget establishments, small angle lamps. Subdued background lighting is required for television viewing and times of relaxation. The room may also be used for meetings. Recessed fixed downlights on a dimmer system will provide a control that allows suitable levels for relaxation, business meetings and room cleaning. Other needs can be provided for by wall lights, table lamps and fixed-task lamps. It should be possible to switch on the bathroom light from the bed. This helps establish a sense of direction should a guest wake up in the middle of the night and not wish to disturb their partner. Illumination should also be considered as regards the position of the telephone. If it is not by the bed, can it be reached easily in the middle of the night? Can this be done without main room lighting?

Wall mirrors in the bedroom can help suggest spaciousness. In the more contemporary interior, the use of a table lamp with perforated metal shade, which casts a pattern on wall and mirror, can dispel feelings of being enclosed. With the restricted area of most hotel rooms, the scale of light fittings must be considered, so that they are in proportion to furniture and room dimensions. For bedrooms which have no exterior windows, frosted glass partitions – uplit to give a glow – or backlit opaque wall panels, provide an illusion of warmth and space. Hotels created from buildings designed for another purpose often have some bedrooms which have no windows. Many guests find this monastic, cell-like, ambience very pleasing.

Restaurant lighting

Hotel restaurants should present appropriate lighting 'scenes' for breakfast, lunch and dinner. Breakfast will require a cheerful but not too bright appearance with a higher level of illumination than dinner. Lunch will need less illumination than breakfast, due to higher levels of natural exterior light. Degrees of presentation and sophistication will vary according to hotel type and clientele. Many hotel restaurants attract customers who are not residents. This repeat business can be a valuable asset to the hotel.

One way of ensuring that lighting reacts to specific hotel needs is with the installation of an 'intelligent', programmable, lighting control system. The use of microprocessors enables smooth mood changes and reactive light management, thus bringing a new dynamic to a wide range of hotel spaces. The microprocessor controls light in areas which have a varied role to play. Once programmed, select scenes can be summoned to serve particular needs and circumstances. Such a system is not relevant to all types of spaces, as not all require changes in lighting and ambience. 'Intelligent' luminaires will also dim or brighten according to exterior light levels and the number of people present within a space.

In most circumstances, a warm light source – which may include candles and a log fire – is preferred for the restaurant environment. If our hotel is an enclave of smoked salmon and dry Martini, then dimmable lamps may be concealed within traditional fittings. In the reinvented nostalgia which may recall *la belle époque*, electronic candles can be used to give a good imitation of the real thing. Such devices are often utilized in plays and musicals as fire regulations do not allow burning candles. Candles, when placed in front of mirrors, create an effect akin to the light-dance of chandeliers.

In a dining space which embraces the world of hard-edged techno-aesthetics, metal halide lamps or 'white' fibre optics can suggest a lifestyle synonymous with avant-garde reality and advanced communications. Spatial experimentation is possible with perforated, wheeled, screen systems perhaps slotted to track-lighting modules, with 'intelligent' luminaires monitoring occupancy patterns of diners. Specific design icons can further suggest a space intended for a clientele who understand the ephemeral language of international design.

The use of simulated daylight, either through decorative stained glass or abstract ceiling panels, can create a harmonized space in what might otherwise have been a gloomy basement restaurant. It should not be forgotten that the quality of light that illuminates the dish is crucial to a full appreciation of the meal. The food should look as good as it tastes. Lamp rendition should be such that the colour of vegetables, fish and meat are if anything slightly enhanced, though this may not be entirely feasible due to the low levels of illumination, especially at dinner. There should be sufficient lighting so that customers can not only see the meal but each other. Electronic or natural candles are useful for dispersing shadowing, which can occur under dimmed overhead lighting.

Between the kitchens and restaurant, there should be a lighting adaptation corridor so that staff can commute between both areas without suffering visual impairment. Kitchens should have bright overall lighting which enables food preparation to take place safely and efficiently. Special care should be taken to avoid glare on food-cutting surfaces and walls which may have metal-faced storage units. Luminaires will generally be recessed fluorescent with electronic ballast to eliminate flicker, and wide-angle prismatic diffusers to provide even, glare-free, lighting. As regards visual adaptation, it may also be necessary to create an acclimatization sector between the restaurant entrance and the lobby or access corridor. This would be relevant especially with the lower light levels appropriate at dinner. There also should be a discreet form of accent lighting to signal the restaurant entrance to guests.

Figure 3.6
Grand Hotel, Southampton, England. De Vere Hotels. Exterior Illumination. Contrast of warm and cool illumination creates an effective visual dynamic within this modern facade. Lighting design by Maurice Brill Lighting Design Limited.

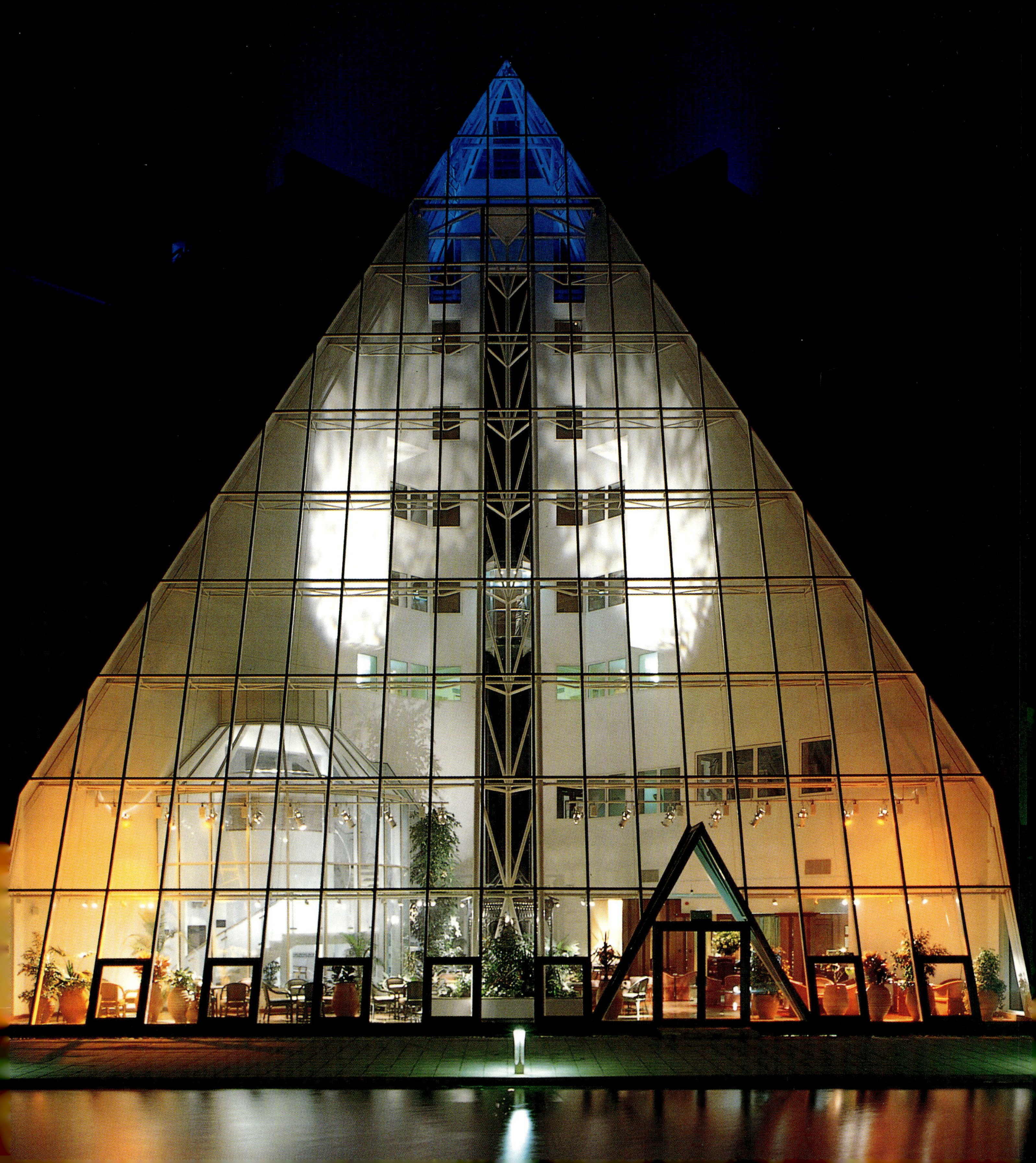

Conference facility lighting

Conference facilities are an essential source of income for the multi-purpose hotel. The volume of business conducted can vary from groups of two dozen to larger parties of several hundred. Illumination should be able to cope with both face-to-face delegate discussion periods and the more focused lecture situation when main hall lights will be dimmed, while directional light illuminates the speaker. Even light on information display boards can either be integral or from ceiling-recessed wallwashers.

There should be provision for conference lighting control from the lecture desk, so that illumination can be dimmed or brightened according to requirements. Audio-visual equipment may also be activated from this position. When main lights are dimmed, individual reading lights should enable delegates to take notes. Depending on the colour scheme of the room, warm or cool fluorescents can establish a particular ambience; for example, 'cool' metal halide lamp sources can suggest an efficient, business-orientated environment. A balance has to be maintained between a lighting solution which is practical, reflecting business needs, and one with a warmer, more human, dimension which encourages delegates to mix and exchange ideas. With this in mind, a section of the conference space could have warm diffuse illumination, while the lecture and presentation zone retains light sources of a cooler appearance. In most conference rooms, a spectacular view from the windows would not be considered a priority. By careful illumination of surface and texture, one would hope to establish a comfortable yet focused ambience, which does not distract from the business at hand. Conference lighting should not in itself be an attention-seeking feature, but should enable those who wish to present themselves to do so with aplomb.

Figure 3.7
Oulton Hall, Leeds, England. Traditional fixtures bring a sense of subdued warmth to this traditional interior. Lighting Design by Maurice Brill Lighting Design Limited.

Lighting in the swimming pool area

Hotel swimming pools come in a plethora of styles and guises, some aglow with daylight, others sunk beneath ground. Electrical lighting for such environments must be ingress-protected from water and condensation. The most important aspect of pool illumination is that the water surface should remain glare-free. This means an even illumination of the pool bottom, without contrast or bright spots. This can be achieved with underwater recessed luminaires in conjunction with overhead, directional, wide-distribution lighting. Diffuse light sources do not provide sufficient penetration of the water surface. Account should be taken of the reflectance value of surrounding architectural surfaces as these can produce glare and water obscuration. Light fittings should be focused carefully.

Natural light should be controlled by use of light shelves, louvres and low-level windows. The unpredictable lux levels generated by the sun can cause dangerous blind spots on a constantly moving water surface. Even north-facing windows will contribute to water glare. Overhead roof windows, which give direct natural light, cause less glare than side windows and provide good visual penetration of the water surface. If specifying tinted fenestration, ensure it does not cause false or unpleasant colouration of the water. Just as customers would be unhappy to eat blue food, they are unhappy swimming in grey or brackish-looking water. In addition, tinted windows will not eliminate glare. Lamp sources suitable for swimming pools are tungsten halogen, fluorescent daylight and high-pressure sodium.

As there is a difference in ambience between a hotel pool and the local municipal baths, a designed lighting environment can be created around the pool. Baths are indelibly connected with ideas of their ancient Roman prototypes, though there is no reason why they should not regress further and become chic paradigms of Babylonian splendour. Too often our municipal pools are no more than plantless, fluorescent-lit water tanks, lined with sanitation tiling. A selection of plants placed near windows can help diffuse direct low-level sunlight.

A hotel pool could afford to introduce a wide range of fabrics and materials; the absorbant materials would help subdue the high acoustic levels associated with such environments. There are many materials developed for use in yachts which would be suited to poolside design. Though the pool itself should be well lit, there is no reason why the periphery could not exhibit an anarchic cynosure of sparkle and style. Though often connected to spartan health-club facilities, hotel pools could celebrate a resonance of luxury and delight in bathing, which goes beyond mere exercise.

The hotel as theatre

Hotel life as a tantalizing and sensual social ritual has in global terms been experiencing a revival, much on a par with the rediscovery of cinema-going, which seemed at one time to be practically extinct. These two modes of sensory marketing have much in common, in that they seek to delight and at times astonish. Just as movies have become exceedingly lavish, so too hotels, whether they be 'traditional grand', 'fantasy', 'designer', or 'retro-moderne', spare no expense in pleasing their chosen clientele. They enable guests to take part in an experience which is about discernment, enjoyment and escape. Contemporary hotels, conceived through the looking-backwards glass, combine tradition with technology and in their miming of the 'new', create a peculiar place, which is perhaps entirely appropriate to the end of this particular century.

4: Art galleries

Caves were the primeval art galleries of our earliest ancestors. Drawings of gazelle and bison took on an iconic magical significance in the flickering light of a night fire. Many of these pictures would not be visible in daylight. The shaman understood the totemic power assumed by such symbols in the shifting chiaroscuro of a wood-fuelled fire. Just as images would come from the flames, so too cloud shadows and shapes would pass continually over the cave drawings, bringing animation and an illusion of space to the confines of an earthbound retreat.

Stone-built churches and cathedrals were the first repositories of the drawn and painted image. Here too, the nuance of light constantly changed and was modulated by stained glass as the sun passed across the sky. Altars, statues and triptychs glowed or smouldered in response to weather and solar elevation. The manner in which holy pictures were revealed was of great concern to the Church. With the gradual discovery of perspective, the painted illusion of space took wing to the vaulted ceilings, so that the stone roof appeared to dissolve into other realms. At night, branches of flickering candle brought their own sombre continuity of light to the assembled throng of angels and saints. There was a reverence within these interiors that in part is still echoed in the softened footfalls and whispered exchanges of our more illustrious art galleries.

The art galleries' roles

The art gallery in its finest expression is a place of retreat, stimulation and contemplation, where human beings can make sense of the myriad symbols that surround them. Some modern galleries are markets of image consumerism and, much as the Medicis displayed privilege and power to fellow cognescenti, the well-informed collector derives venial satisfaction from the visions of past and present. Inevitably, the commercial art gallery reassures the collector that what they are purchasing is of some intrinsic worth. In the national and public collections, this sense of artistic uncertainty is not so prevalent. If the experts are unsure of a work's provenance, it is not displayed until such times as the dilemma is resolved. This dichotomy of motivation generates a change of emphasis between the commercial and public display of art.

Figure 4.1
Thyssen Museum, Madrid, Spain.
Architect Rafael Moneo. Lighting
design by Ove Arup & Partners.
Erco Lighting Limited.

The illumination in a commercial gallery is usually crisp without being too cool or severe. There is a strategy to attract clients who will buy, while gently discouraging idle browsers and cultural tyre-kickers. To this end, luminous precision is demanded in the window display area, with slightly warmer lighting on the sales desk within.

There are variations on this theme depending on whether the art displayed is modern or traditional. In the traditional gallery, too cool an ambience could deter the more conservative buyers from entering, as their conception of quality may be connected with a low, warm light suggesting historical elegance. To avoid a disjunction between art and surroundings, care must be taken in providing a colour temperature of lamp which is appropriate to both paintings and room architecture.

Conservation lighting

Figure 4.2
The Galician Centre for Contemporary Art, Santiago de Compostela. Architect Alvaro Siza. Cool illumination provides a contemplative neutral backgound to a work by Richard Long. Erco Lighting Limited.

Public art galleries which exhibit historic paintings are aware of the damage caused by light. In order to ensure conservation, the ideal condition would be one of absolute darkness. This of course is not practical. Paintings which are not on display should be stored in a temperature-controlled environment, where there is no possibility of light damage. In the exhibition space, a compromise must be made between providing enough light for comfortable viewing, while carefully limiting the excess lux levels which cause surface and substrate deterioration. Historic oil paintings should not be subject to more than 200 lux, while watercolours, drawings and manuscripts may receive a maximum illuminance of 50 lux. As well as controlling ultraviolet and visible light, these lux levels help to reduce the amount of heat generated by the lamp. A crucial factor when dealing with fragile surfaces is the length of time per annum that the exhibit is lit.

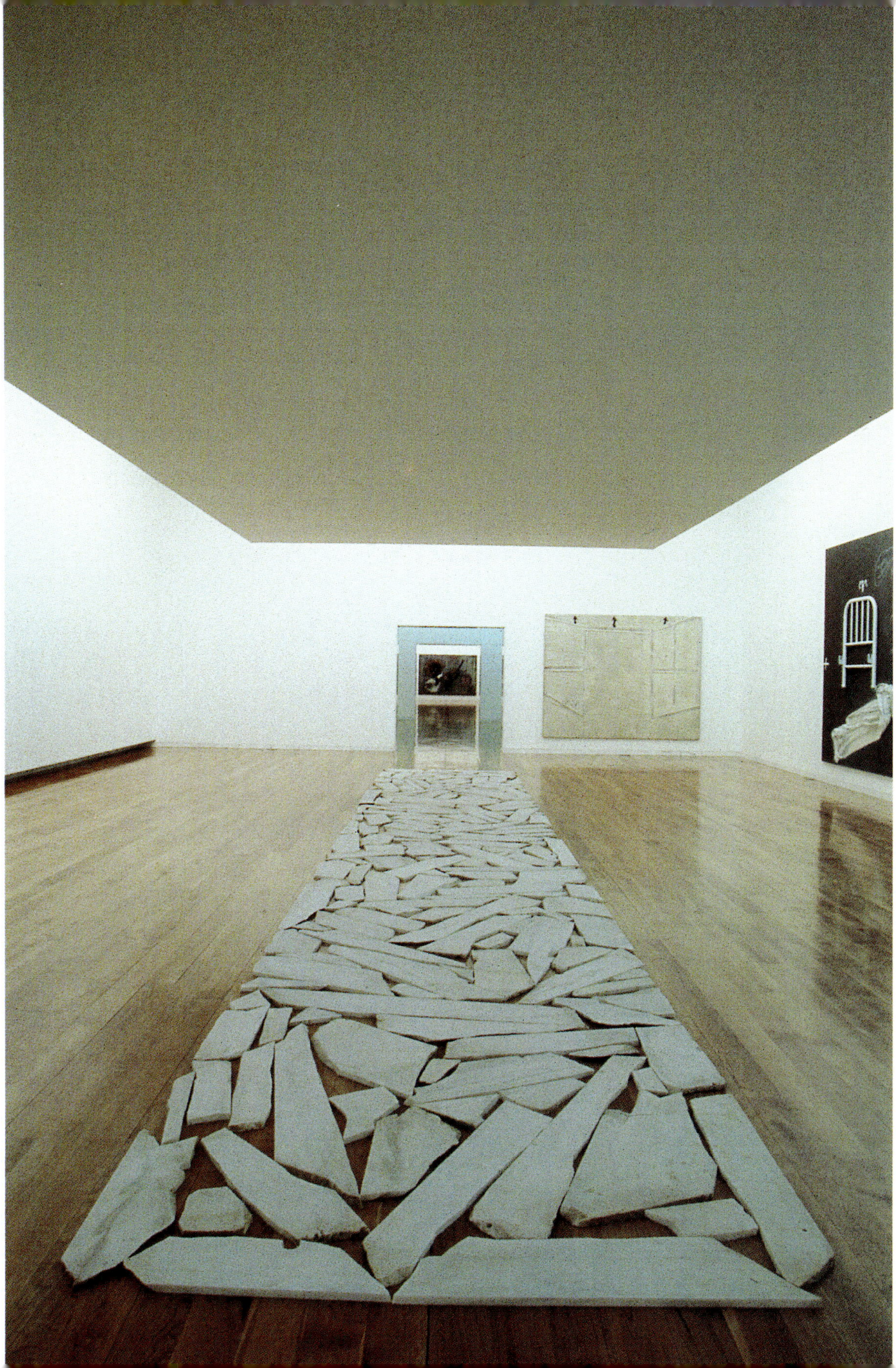

Some pigments used in oil and tempera paintings are eroded by light and it may be necessary to exhibit such works on a limited basis. Therefore paintings of a similar age may, due to differences in paint composition, be shown for disparate periods. All daylight produces ultraviolet rays and protective measures should be taken accordingly. This usually consists of UV filters on windows in the form of film, or on picture glazing, or over the actual light source. Most fluorescent lights will require UV filters.

In galleries which have 50-lux illumination limits, consideration should be given to connecting corridors between areas of high and low light. Too high a level of light in corridors or adjacent rooms will cause visual adaptation problems. In spaces where watercolours are exhibited under levels of 50 lux, daylight penetration would prove to be impractical. In some cases, diffuse natural light through drawn blinds may be acceptable in conjunction with artificial sources. Solar illumination varies dramatically and is expensive to control.

Architects generally prefer natural light for picture viewing, but this predilection is not encouraged by conservationists, who prefer a source which can be controlled. Reflected daylight can be used in excess of 200 lux if there is an annual restriction of lux-hours exposure. Light sensors monitor lux-hours exposure per annum.

Internal temperature is measured and louvres on the roof adjusted, each element recognized by the computerized control system, which is specially calibrated to prevent excess ingress of natural light. Such a pact with daylight necessitates that all sun blinds be closed outside public viewing times. If daylight is permissible, dimmable warm fluorescents or wallwashers of good colour rendition are required on days when there is little natural light.

Figure 4.3
The Richelieu Wing in the Grand Louvre. Careful accent lighting emphasizes form and surface texture. A marble bust of Jules Hardouin Mansart, who built Versailles, Les Invalides, and the Marly-le-Roy chateau for Louis XIV. Erco Lighting Limited.

Lighting for optimum viewing

Windows in corridors connecting high and low light areas should have blinds to prevent excessive glare. Glare from windows or unshielded lamp sources affects our ability to discern colour and detail when viewing paintings. The problem of reflection caused by badly focused spotlights can be exaggerated by the varnished surface of large canvases which extend high up the wall. Anti-dazzle louvres and careful alignment of spotlights can help to reduce surface light on canvases.

The problem of glare and reflection is not so apparent at lower levels of illumination. If 50 lux or lower is required on exhibits and this is evenly applied to all surfaces of the room, then the resultant similarity of tone will adversely affect our appreciation of the space. A subtle dipping of lux levels between paintings helps to introduce visual vitality to the room.

This variation of light level suggests that there is more light on exhibits than there actually is. Caution should be exercised in exaggerating the lux dip between paintings, as this can cause a 'searchlight effect' which negates the overall integrity of the viewing space. If the 'drop off' is too great, the eye requires a longer time to adapt between exhibits. At low light levels, a lamp of warm colour temperature is more acceptable than a lamp of cool temperature as a cool light can have a psychologically depressing effect on viewers. This, allied with the probable exclusion of sunlight, would not provide a comfortable environment for art. Tungsten lamps give a warm illumination for low-lux viewing and emit little in the way of ultraviolet radiation. In some situations fibre optics will eliminate the problems of direct heat and radiation, while giving sufficient light for watercolours and manuscripts. Fibre optics are also suitable for display cases.

The choice of warm or cool light for paintings will depend to some extent on the historical ambience which the gallery hopes to create. This is a subjective matter and the way such choices are approached will vary considerably. The paintings of Turner, for instance, might respond to a warmer environment in sympathy with their worlds of fire, smoke and sunlight diffused through evening clouds.

Conversely, the work of Vermeer and Van Eyck might benefit from a cooler presentation, where the northern light and attention to detail would be revealed. Whether warm or cool, a light providing good colour discrimination is essential. A room devoted to paintings from a particular period of time will usually be lit with lamps of identical colour rendition. They should all be sourced from the same manufacturer.

A combination of diffuse and directional light brings a range of lighting options to spaces where exhibitions will only be on display for limited periods of time. Surface illuminance of paintings should be regularly checked with a hand-held light meter. Excess use of strong, diffuse, light causes texture and pigment detail to be drained from the canvas. With works by such artists as Rembrandt or Soutine, where the surface quality of paint is important, this would affect our visual understanding of the work.

The reflectivity or light absorption of wall surfaces will alter not only the paintings but the room in which they are hung. Dark historic paintings on a light wall will appear isolated and difficult to relate one to the other unless a device is found to link them visually. A mid-tone wall of maroon or damask red, warm ochre or bottle green will infuse a unifying element to room and paintings, and in many cases enhance old pigments.

It is acceptable to use rich colours on gallery walls where traditional paintings are being displayed. Red, in particular, when used on walls creates a rich contrast with ornate gilt frames. Blue is generally not a colour that draws in the viewer, and is only used effectively when a particular interior architectural statement is being made, for instance as a contrast to warmer hues of adjoining galleries.

Strong wall colouration is not so prevalent in galleries showing modern art, where a light or neutral background is usually specified. If the lighting for such a scheme is badly handled, it can convey a latent mood of mediocrity, where paint and wall-surface imperfections are evident. Well conceived, this neutrality allows attention to be focused on exhibits which are often huge in scale and destined for the corporate art market, where there are gaps large enough to accommodate them. The modern commercial gallery must interface with businesses in a way which persuades them that they have an appetite for images, and that art is essential to their existence. The modern gallery space must therefore be sharp and vibrant, a brief dialogue which bridges the gap between business and art. The lighting will usually be cool with fluorescent, low-voltage tungsten halogen and metal halide lamps being recessed or visually concealed within the gallery space.

In conservation rooms, where detailed examination and repair work is undertaken, higher light levels are permissible, as exhibits are present for limited periods of time. There should, however, be a darkened holding compartment so that fragile work is only in full illumination when being examined and restored. Up to 1000 lux is allowed, though the lamp should not be one that emits excess heat. The conservation lamp source should have the same colour rendition as the lamps under which the work will be exhibited. This prevents potential mismatch of retouch paint which can become visible under the light of a different lamp.

At night, when the gallery is closed to the public, illumination of the paintings should be avoided. A restricted amount of lighting is necessary for security. With infra-red surveillance detectors and cameras, it is possible for interior sections of the gallery to remain in darkness. Illumination of the façade and immediate perimeter area should be sufficient for security, though care should be taken that it is not so bright as to cause marked visual contrast thus affecting the night sight of guards.

Lighting the contemporary art gallery

There is an open quality in a gallery of contemporary art, which makes it quite different from one which houses the treasures of the past. The element which causes this palpable disjunction is the expansive quality of natural sunlight, allowed to flood through the voids, upon art freed from the restraints of conservation.

Contemporary galleries often house travelling collections or installations which will only be in place for a matter of weeks. Natural light provides a connection between the world outside and the diversity of the artwork within. With changes in weather, the atmosphere of the space is constantly adjusting, allowing exhibits to manifest differently from day to day. Depending on tonality and texture of walls, illumination will be diffused, directed or allowed to flow free. In this way, art and architecture are united in a reciprocal relationship with the medium that reveals them. Artificial light will be required during the later hours of winter months. Even during periods of low light, the experience of windows opening outwards brings a sense of reassurance to those within the building.

This is a different world from the enclosed sarcophagus of an historic collection. Walls, illuminated and made of concrete, travertine, glass or wood, infuse the voids they create with a quality particular to that material. This transmutation does not occur in an environment lit entirely by electric light. The moving, translucent warmth of sunlight is what provides the magic animation of space.

Many sculptures and built installations are conceived out of doors and can appear moribund when lit by artificial sources, just as driftwood can lose its charm once indoors. Although designers talk reassuringly about controlled daylight, it remains difficult to control,

Figure 4.4
Sainsbury Wing, National Gallery, London. Photography Dennis Gilbert. Erco Lighting Limited.

yet when given some rein imparts that edge of vibrancy to interior space. Once it is completely controlled, you might almost be as well-off with simulated daylight. In the traditional gallery, where daylight is strictly controlled, there is a sense of unreality to some of the objects which are only viewed under the constant illumination of electric sources. In a new gallery of modern art, there would be a strong case for allowing daylight exclusive access to one zone of the interior, even though all that was exhibited consisted of carved stone, iron and glass. A variety of natural light sources and a collection of ceremonial artefacts might provide a anarchic centrifuge between ancient symbolism and the latest organic explorations of stone sculpture.

A gallery displaying work painted in the last fifty years will usually trace a sequential route which follows the development of particular artists. In this situation, a matter of only a few years may separate one gallery space from another. It is interesting how recent periods of apparent risk and artistic daring can acquire a whiff of parody when viewed together in the same space. How readily we now recognize a clutch of paintings representing commercial gallery offerings from the late 1960s. Recent work of this type can share an overall illumination level with the exhibition space, without there being a need for a drop in lux levels between paintings. (Passing years bring to autocar spray paint a distressed authenticity to rival the surface of a badly restored old master, and achieved in a fraction of the time.)

Lighting can help to lead visitors along a predetermined route which may help to aid their comprehension of artistic change and development. In the more contemporary gallery, there will be no need for visual adaptation areas. Most spaces should present a happy compromise between natural and artificial light. Subtle accent lighting can guide visitors to the correct doorway and, where there is a choice of adjacent entrances, a slightly higher lux level will indicate the preferred route. A fairly wide range of lighting equipment, including tracked adjustable spots, recessed fluorescents and wallwashers all with UV filters and lamps of good colour rendition, will be suitable for this type of gallery. Care should be taken to avoid hanging paintings where they will receive direct sunlight. Even paintings of fifteen to twenty years old should be protected from the direct effect of sunlight, for these may well be the masterworks of the future.

Light levels will be higher than in a gallery housing historic art, therefore care should be taken to reduce glare caused by a lighter decor and polished-wood floor surfaces. This will require accurate focusing and alignment of tracked spots, which should also be fitted with anti-glare louvres. There can be a problem in using too much diffuse light within the pale mid-toned environment of a modern gallery. A visual 'white-out' can occur, which flattens surface texture, drains colour and confuses our perception of structural dimensions. In some minimal exhibitions, this technique is deliberately employed to create a soft nebulous miasma alienated in atmosphere from the rest of the building. However, this is not useful when concentration, visual awareness and appreciation of complex themes are required.

Other gallery spaces

The audio-visual room

The gallery which houses a national collection will be unable to display all of its works at one time. There will be days when it is not convenient to hold educational talks for schools or societies within the main gallery area. An audio-visual lecture room can provide a separate space where relevant information can be conveyed without distraction. Such a room also provides an opportunity for visitors to view work which is not displayed by means of slides, photographs or PC screens. The first illumination criterion is that lecturer and audience should be able to see each other without difficulty. If using directional downlights, the beam should be of a wide distribution in order to reduce facial shadows. The transfiguring charm of light is somewhat reduced if the audience look like extras from *Night of the Living Dead*. Interior designers will sometimes promulgate a theory of compromise when confronting audio-visual spaces, arguing that people are there to view the images on display and not each other. The way that the audience members perceive each other and also the speaker will affect their response to the material presented. The ubiquitous fluorescent is anathema to many young designers who view it as a moribund traducement of draughty classrooms and abandoned lock-up garages. This subjective aversion blinds them to the recent advances in fluorescent technology which can make them ideal for a variety of teaching and lecture situations. Dimmed louvred fluorescents with wide light distribution can offer a responsive solution to audio-visual needs.

The recent generation of lamps will give good colour rendition of facial tones and the addition of diffuse wall lights helps to produce a degree of facial modelling. Wide-beam directional downlights would be acceptable for illuminating lecture boards. Here, lux levels can be slightly higher than on the audience. The lecturer should have hands-on control of the room's various dimming options.

Computer technology can now present quality reproduction of historic and modern paintings. Where monitors are used for this purpose, lighting should utilize low-brightness, glare-reduction louvres to prevent reflection and veiling of the display screens. Curtains should be capable of darkening the room so that there is no leakage of external light. Lamps which generate large amounts of heat should be avoided if possible. There is nothing more distracting than trying to listen to a lecture in an overheated room where the curtains, though closed, let in chinks of light.

The gallery café

The gallery café or restaurant is usually a composite which conforms to the language of the building. It could often be another gallery space, packed with tables and chairs designed for people of indeterminate taste. Seldom is there an attempt to celebrate a sense of occasion, though this is where reactions should be discussed and opinions aired. Interior designers are wary of interrupting the tone of a gallery experience, and thus the café is often somewhat lacking in heart or character, whether it be a traditional compendium of stained oak or bentwood Bauhaus.

With succinct illumination it is possible to create a intriguing change of mood without breaching the building's overall architectural integrity. It may well have been that when constructed the building itself was somewhat at odds with the art it was designed to house. However, given the passing of years, even the most recidivist examples of whale-bone brutalism mellow and find their rightful place as servants of art. With interior lighting we are merely altering the way that spaces are perceived, especially in relation to their use, and a café is a space quite distinct from a gallery. The café, even if it boasts a linoleum floor masquerading as mosaic, should stimulate conversation and inject an irreverent insouciance to the otherwise rather contemplative business of art. Galleries are usually quiet places, but the café of a good art gallery should be positively buzzing with excitement and debate.

A lighting designer could devise a strategy by which the distinctive use of shadowing suggests warmth and safety, yet retains enough dynamic to stimulate conversation. This could be achieved without diverging from the architectural language employed throughout the building. In the café, daylight establishes contact with the outer world and may be reassuring after passing through a series of electrically-lit, windowless, gallery spaces.

Blinds should deflect direct sunlight and lamps utilized to give good colour rendition for both skin tones and food. The use of light and degrees of shadow, either with tracked spots or recessed adjustable downlights, brings a sense of depth and chromatic variation not present in most art-gallery spaces, where the main task is illuminating the exhibits upon the walls.

In winter months individual table lamps can bring a welcoming glow to the café area. Our understanding of a café is synonymous with a sense of safety and the free exchange of ideas. It is possible to create an emblematic retreat where people do not feel they are being hurried to leave, as in fast-food outlets, or overtly fussed over as in grander establishments. Light can change a space without the space having to change. For the culturally dispossessed who find gallery visits a penitential ordeal, the least we can offer is a pleasant café in which they can recover their composure.

Lighting for gallery security

Art galleries are an avatar of visual avarice, housing more in the way of transportable desirables than a clique of Swiss banks. Security is a priority. Video cameras at the entrance should record all persons entering and leaving the building. This establishes a long-term record of who has visited the collections and in what circumstances.

Even though a robbery may occur at night, videotape will give police a basis for identifying potential suspects, as such crimes are usually planned through a series of daytime visits. It is essential that lighting in the security camera zone scanning the entrance should be sufficient to insure a positive identification of people entering and leaving the building. Narrow-beam directional downlights are to be avoided as they create distinct shadowing on facial features and blurred video images. Background lighting should not be strong as the contrast will induce a silhouette effect and also should not exceed three times that of the light on the visitor. If possible there should be natural colouration of facial features and if cameras are positioned at the entrance, lamps should be of similar colour temperature to daylight. Louvred overhead fluorescents with wide-beam distribution, in conjunction with uplights mounted to either side, will provide good facial modelling.

A gallery is responsible for preventing the theft of works from collections and ensuring that those works which remain are protected from the negative influence of excess light and chemical degradation. There will always be a dichotomy of interest between the public's desire to see paintings and the gallery's need to protect them from decay. Photodeterioration depends not only on light but on the atmosphere and temperature of the storage areas. Bad conservation can be as guilty of theft as the midnight machinations of determined robbers. Once started, the wheels of chemical deterioration continue turning, even in darkness.

Figure 4.5
National Museum of Natural History, Paris. Architect Jules Andre. Lighting settings, Rene Allio, Paris. Beneath an artificial sky discreet layering of ambient and accent illumination leads visitors through a vast array of exhibits. Erco Lighting Limited.

Bringing the gallery to life

Natural light is an essential adjunct to art: it is light which reveals the essence of the artist's experience, for language alone is quite inadequate even when confronted with a canvas which displays a strong literal premise. Although artists usually describe their colours as individual hues, the lighting designer's spectrum extends in a continuous band, and it is this cuirass of orchestrated continuity that provides a stable gallery environment, for the exploration of the imagination is an intense experience, encompassing the worlds of both angels and demons. To some, the traditional art gallery resembles a tomb of the ancestors by which we can return magically to fixed points in time. There are pharaonic undertones in a nation's desire to carefully preserve the best of the past, as if this were an anchor lodged against uncertainties to come. It is this echo of the tomb or the church which accounts in part for the hushed gallery tone, which habitually alienates sections of the public and which gallery administrators have tried with varying success to counteract. So many stratagems have completely failed.

Perhaps like those who refuse to drink Guinness, there are many among us who just do not care what is good for them. It is interesting that few articles or papers on art galleries actually mention or discuss the role of the visiting public. It is as if, after the experts have had their say, the public are considered as a bolted-on afterthought. It is generally agreed that city galleries should actively engage the community, rather than remain static repositories of historic artefacts. How many galleries have an ongoing relationship with a substantial percentage of city schools, where a visit is a regular occurrence rather than a rare expedition? The running costs of gallery buildings are high and the final decision on whether a gallery remains active or passive rests on its director's skill and the input of politicians who arrange the funding.

Lighting design is a recent factor in the gallery equation and extends beyond the concerns of conservation, providing an opportunity to improve the experience of looking at art. Natural light, which, like art, emanates from the past, yet is seen to exist in the present, provides a bridge for perception from which the heights and depths of the human spirit are briefly glimpsed.

5: Exteriors

From firelight to electricity: exterior lighting and imagination

Prior to the invention of the gas lamp, a building was seldom visible at night, unless it was burning, or bathed in moonlight. The beauty of moonlight so captured the public's imagination that attempts were made to mimic this effect in theatre productions, even before adequate directional light sources had been devised. Moonlight would have been a known urban phenomenon, before the days of electricity, industrialization and night-time light pollution of the skies. Today, there are city dwellers who have seldom seen moonlight.

In nineteenth-century Germany, the theatre designer Angelo Quaglio thrilled audiences with lighting effects on stage sets of painted towers and city walls, created for the world premieres of Wagner's *Tristan und Isolde*, *Das Rheingold* and *Die Walküre*. For example, moonlight, or the presence of supernatural beings, could provide the explanation for a building's night-time illumination. In these productions, architectural elements were strikingly illuminated before the introduction of electricity, but the use of combustible materials as light sources meant that many theatres burned to the ground.

The idea of supernatural intervention generating light was developed by romantic painters such as Arnold Bocklin, whose 'Island of the Dead' (1880) was suffused in an amber glow that appears to emanate from the rocks and buildings themselves. The painting imparts a mysterious sense of unease as tall trees shift in the wind, but the sky is dark, displaying no trace of solar or lunar light.

In England, Victorian artist Atkinson Grimshaw (1836–93) painted moated manor houses transfixed by the quietude of moonlight. Many of these paintings, with their autumnal undertones of ebony and decayed fuschia, have a chill, hallucinogenic, quality akin to a land of disturbed dreams. People have always been fascinated by gardens and architecture made visible at night, yet for centuries, when the sun set, a world beyond the windows was denied us, and retreated to the realm of childhood fears.

During the 1920s, directors such as Erich von Stroheim used powerful kleig lights for filming night scenes of gigantic exterior sets of the Casino at Monte Carlo and the Café de Paris. The spectacular nocturnal illumination of Grauman's Chinese Theater for gala film premieres brought packed crowds to Hollywood Boulevard, where coconut palms, ornate lanterns and a probing palette of searchlights lent a scintillating aura to this bizarre building before which movie stars dismounted from their limousines. The use of 880,000-candle-power searchlights for outdoor extravaganzas was pioneered by Otto K. Olesen, whose illuminations of not only film premieres but the Hollywood Bowl, the Golden Gate Bridge and many other buildings and festive occasions fascinated thousands of onlookers. The public wanted to be astonished.

Figure 5.1
Schillerbad Brewery, Ludenscheid, Germany. Architects Altenheimer & Wilde, Ludenscheid. Former art nouveau swimming pool converted to brewery, bar and restaurant. Erco Lighting Limited.

This desire to capture the public imagination crossed into billboard advertising, where animated electric images were incorporated with flashing letters and special effects. The famous Camel cigarette billboard of Times Square, in which a man puffed rings of illuminated smoke, was erected in 1941 and captivated passing crowds until it was switched off in 1967. It was from these high-octane beginnings that exterior lighting gradually became accepted as an enhancing feature of the nightscape, rather than as a spectacular one-off celebration.

The exterior illumination of prestigious homes in Beverly Hills and BelAir was an extension of the Grauman's 'look at me' scenario, allied with a need for effective security. The rolling electric arteries of the freeway could by the 1950s be seen from the top of the Santa Monica hills. The automobile as a fraction of this kinetic sculpture became not only a symbol of social standing, but an essential tool of the daily commute. It provided an extension of the well-illuminated house, which not only protects but proclaims the success of its owner. Tail fins and myriad indicator lights were developed to a point far beyond the functional.

Down to earth: security lighting and other practical considerations

Most homeowners illuminate their property as a security measure. This is an important consideration. No residence can be made one hundred per cent secure, but several devices are available which greatly reduce the risk of robbery: PIR infra-red detectors respond to changes in surface termperature within the beam, which will then trigger lights, CCTV or an alarm. Most break-ins are done by passing opportunists, who if faced with counter-measures would swiftly retreat. A low level of illumination around a building will act as a deterrent and can be used in conjunction with PIR detectors linked to an alarm. A master switch to activate additional security lighting can be located within the house. The usual place for this is either in the kitchen area or the master bedroom.

The task of exterior lighting should be approached from a practical standpoint. Shielded functional lighting can create safe access areas. These may be to the rear or sides of the house. Care should be taken that there is no spillage of light into neighbours' property. The route between garage and residence could be lit through the triggering of PIR detectors. Shielded bulkhead fittings are suitable for the rear of premises, insuring that those leaving by automobile will not be dazzled. Stake-mounted mushroom luminaires are useful for lighting low plantings and can double to delineate the edge of pathways. Light fixtures should be sited clear of lawn sprinklers, which can rust metal casings. Leaves, dirt, animals and insects can effectively reduce the life of exterior lighting systems. When using recessed well lights, or direct burial lights, which are sunk into the ground or paving, the water drainage flow of the surrounding topography should be studied and taken into account.

Lamps suitable for exterior illumination include tungsten halogen, mercury vapour, metal halide and high-pressure sodium. The lamp chosen will depend on the area to be lit, nature of climate and acceptable electrical costs. Specifying lamps which are dimmable, such as tungsten halogen, provides visual flexibility and power cost savings. If a client is spending money on exterior illumination there is an argument for using a source such as tungsten halogen, which may be more expensive to operate than high-pressure sodium, but gives improved luminous quality and can be used with lighting control and dimming systems. These can alter illumination levels automatically so that several different 'scenes' would be possible in one evening. In other situations, lighting could be switched off. For instance, full illumination of the space may not be required after midnight.

The world outside: garden lighting

During daylight, there is a visual link between the rooms and windows of a house and the garden beyond. Sympathetic exterior lighting can help re-establish the idea of the garden as an extension of the residential environment. Our attempts to bring order to nature can range from the sublime abstractions of the Zen garden to the ornate Baroque splendour of Versailles or Vaux-Le-Vicomte, which was created by Andre Le Notre (1613–1700). In the Tuscan gardens of the Villa Marlia, Paganini once played in a torchlit enclosure of sculpted yew hedges, to an awestruck assembly of Franco-Italian aristocrats, who in turn were observed by the shadowed statues of Saturn, Jupiter and Pomona, attended by carved stone harlequins and mountebanks. The gardens of the past were lit by torches or by brief bursts of fireworks.

The notion of a garden representing a symbolic microcosm of man and the universe has been incipient since the seventeenth century. French gardens conjured up a harmonious flow of perspectives which integrated ideas of visible and mythological worlds. This extraordinary encapsulation of timelessness, of an invisible geometry evoked by gardens such as Vaux-Le-Vicomte does on a certain level conjoin with the meditative austerity of the Zen experience.

A designer sculpting with light must be receptive to the qualities a garden possesses. At the same time, it is our restrictions on nature that produce a beautiful garden and this fundamental process guides our approach to the lighting, so that we plan perception in a rational and precise manner. The prerequisites suggest an ambivalent affiliation between the pragmatic needs of utilizing a viable lighting programme and considerations both unconventional and experimental which examine the subtle and complex possibilities of a garden, for instance, as an astrological anagram. The space does not have to be a cornucopia of box parterre, pergolas and bowered fountains for the illumination to create a sense of wonder and occasion.

Lighting trees and shrubs

Different species of shrubs and trees will require specific lighting techniques, depending on their size, colouration and location. Lighting tactics should take into account the fact that plants and trees increase in size. In winter, deciduous trees take on a different appearance, and this seasonal reconfiguration should be allowed for. The textural quality of tree bark provides a useful vertical accent which can be illuminated both summer and winter.

Flowers should be lit with lamps of good colour rendition, such as tungsten halogen, which are available in a wide range of wattage and beam patterns. Metal halide lamps also give good colour rendition and are suitable for both plant life and building façades. They have a longer lamp life than tungsten halogen but in some situations may appear too cool. High-pressure sodium has a gold-tinted light which is suitable for stone façades or garden statues. It has an excellent lamp life and can be used in conjunction with metal halide if suitable colour rendition for shrubs or trees is required. Mercury vapour lamps give a crisp blue-green light which can be used to accentuate fountains or foliage.

An intriguing garden is greater than the sum of its parts, and its illumination requires finesse, so that surfaces are brushed with light rather than starkly revealed. You will, of course, find the occasional client who wants their patio lit as if to attract flying saucers. On such occasions, a certain amount of tact and diplomacy is required. Attempts to accent too many areas will result in visual overkill. Areas of interest should be linked by low ambient light, so that there is a sense of connection throughout the scheme.

Trees with a generous spreading canopy can be lit from directly beneath so that the leaves are etched in an intricate pattern against the sky. Trees adjacent to a wall or façade can be side-lit with a single stake-mounted spotlight, creating a dramatic contrast of texture and spatial dynamics. Carefully angled, the beam will cast foliage patterns upon the wall. The shape, size and leaf colour of trees should be taken into account when deciding what lamps are required and their position. It is better to illuminate with less rather than more. The main viewing angle should also be considered, whether it be from a window or terrace. Accent lighting can be used to edge sculpture, steps, tree bark, fountains and archways. It can act as a visual signpost, and a way of leading the eye through a scheme.

The technique of silhouetting trees was effectively conveyed during the 1660s in the magnificent paintings of Claud Lorraine, who accentuated his darkened foliage with a rich transcendent aura of golden light. Here were unprecedented classical gardens fit for both mortals and gods. Eschewing oil paint for electricity, the effect can be recreated by placing stake-mounted luminaires behind a suitable grouping of trees. Consider the main viewing angle. The beams should graze and slightly intersect the upper lip of the canopy, thus absorbing direct light, and reducing the danger of contributing to night-sky pollution. This technique can be used to good effect if trees are silhouetted against a wall, so that their mass and movement is revealed. Shadows can be used creatively within the overall exterior setting.

Figure 5.2
Department of Works Head Office, Witten, Germany. Architect Prof. Jorg Friedrich & Partner. Translucent fenestration allows effective contrast between warm interior and cool exterior illumination. Erco Lighting Limited.

Lighting other garden features

Decorative luminaires will include lanterns, twinkle lights, neon, fibre optics and candles. They can help bring a sense of enclosure to an exterior seating area. This may in turn encompass an ornamental pool or piece of sculpture. If using colour filters with exterior luminaires, it is visually more effective to explore the cooler blue-green end of the spectrum, rather than employ a palette of reds and yellows which will give leaves a grey or brown appearance. On a light-coloured terrace some pinks or yellows may be acceptable though there is a danger of creating a village disco look. In a house with large panoramic windows, it is possible to link exterior colours with wall colours visible within the house. In this way the house extends on to the terrace or patio.

The reverse aspect of this idea of sculpting with light would be the solar garden, enclosed in a skeletal extension of the residence, carved by shadows which redesign themselves during the course of the day in the form of mobile geometry. Such spaces could only operate in countries with hot climates. Either way, the garden is an emblem of impermanence, whether the shadows are those cast by solar rays or spiked spots.

In warm climates, the garden is more usually perceived as an extension of the house. Sunlight will fill the garden for a large proportion of the year. With the appropriate lighting technique, moonlight can be guaranteed for every night of the year. This is produced by placing lamps with medium to wide downward reflectors high within the trees. Beams shine down through the branches, causing a dappled pattern upon the ground. Lamps used should give a blue-white illumination. Sources suited to producing a moonlit effect are metal halide and mercury vapour. This cool ambience can be contrasted with the decorative warmth of incandescent or candle lanterns scattered around a terrace or patio.

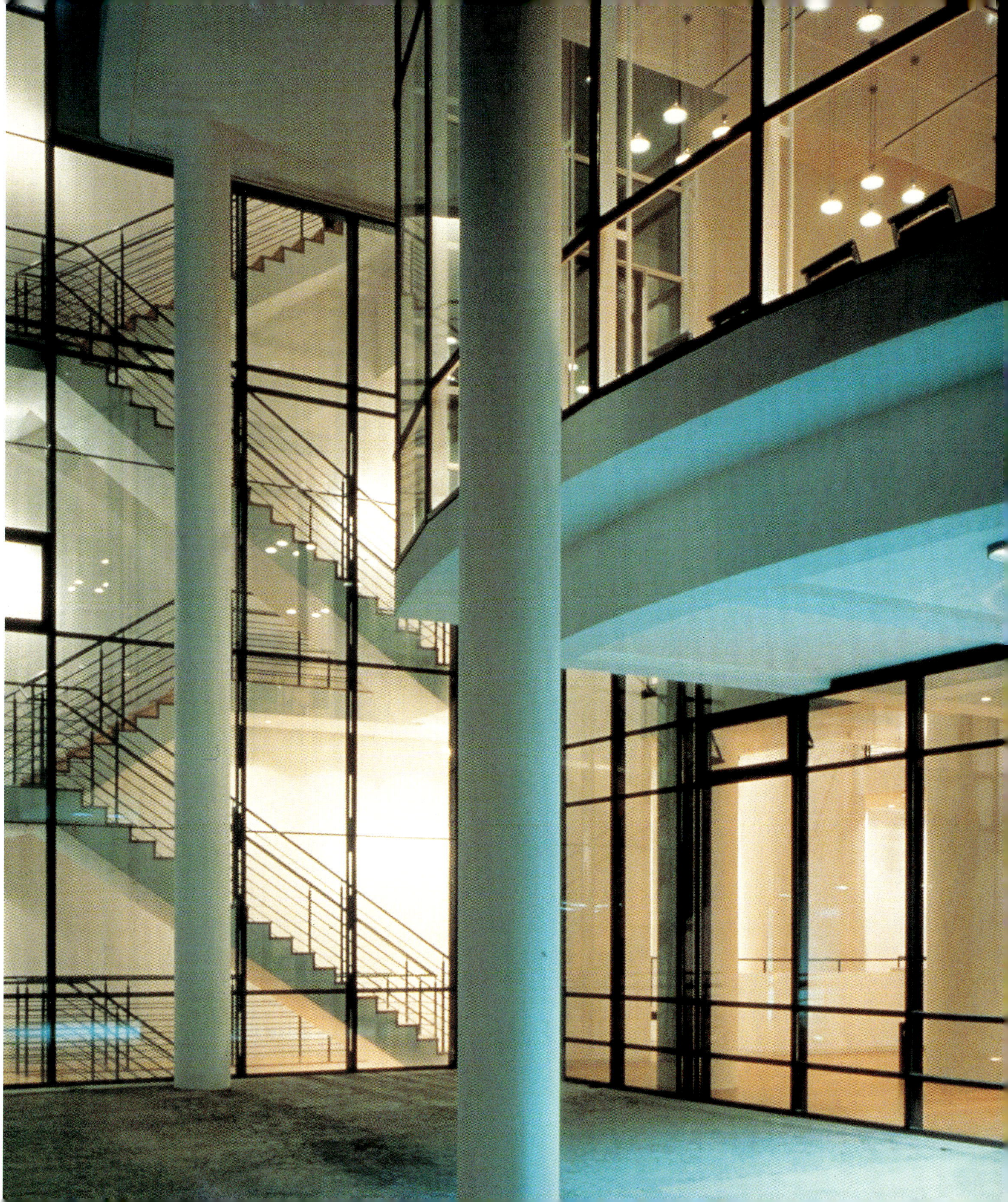

Highlighting architectural detail

To show architectural detail, façades should be illuminated by a number of medium- to narrow-beam luminaires, rather than a lesser quantity of powerful wide-distribution floods. Strong floods will wipe out surface detailing, so that from a distance the building may resemble a cardboard fold-and-build heritage toy. Where possible, fixtures should be concealed so that they are not visible during the day. Lamp sources such as low-pressure sodium should not be used for façade illumination, for though they are a most economical source, the quality of light produced is a grim orange fit only for the nether regions of Hell. Façades should not be lit directly on to the dominant viewing angle as this flattens form.

Lighting can also reveal imperfections, so time should be spent checking surface quality. The tonality of façade material will affect the amount of light required to illuminate it. Metal halide lamps help to accentuate cooler, bluish, stone, such as slate, flint or granite, while high-pressure sodium brings out the warmth of brick or sandstone.

Open-sided fibre optics can be used to draw with light around architectural features, swimming pools or stairways. As directional light points, fibre optics can be positioned like stars beneath a canopy or pergola. Set beneath a handrail, fibre optics in their directional or open-sided mode can give subtle indication of an edge or boundary.

A stronger form of delineation could be achieved using neon or cold cathode, rather than open-sided fibre optics. Neon can be utilized as a form of connective sequencing to accent the linear quality of buildings or describe related spaces. It is also useful in situations where lamp changing is difficult. Both neon and cold cathode have extended lamp life. Their decorative application is most usually commercial.

As I have mentioned earlier, the needs of an attractive façade and security lighting are not irreconcilable. Façade lighting should reveal a building's salient features in a compelling manner, without wiping out either form or surface detail. Strong exterior directional lighting is usually only suitable for prison perimeters and football grounds. Light should caress façades, sensitizing surface textures and revealing architectural nuances which may not be immediately appearent by daylight. A carefully illuminated building can look extraordinary at night, when compared with its somewhat muted incarnation by day. It will not only be a question of whether warm or cool light sources are used, but the building's location in relation to the surrounding space. A tall building within narrow winding streets will present quite different challenges than a small country house set within fields. It is generally agreed that floodlights should be concealed but not in such a way that they might present an unexpected hazard to wandering persons.

In a town, fixtures may be mounted on the walls of a building, if there is no suitable location at street level. These can be painted to blend with the façade. Spill light from surrounding buildings should be considered. It is not necessary to light an entire façade or roof in order to make a pertinent visual statement. Just as a hat can inform us of a person's occupation, so too the roof protects and comments on the structure below. Whether this be the creamy extravagance of Palm Beach's villa Mar-a-Lago, where sweeping roofscapes imitate the ocean waves, or the chill lunar geometry of a remote state correctional facility, there is a cryptic luminous language designating structures in darkness.

An illuminated roof area should suggest an extension of the façade, while making succinct comment on the character of the architecture. Roof illumination usually culminates in some significant detail, which may not be the highest point of the building. It could be a turret or decorative balcony which creates a visual connection with the façade or entrance. To scatter the roof with a plethora of spotlighted details would be like covering a birthday cake in cherries.

Figure 5.3
Invalides Square, Paris. LED luminaires provide an elegant curve to the city nightscape. Lighting Design LEC Lyon, France.

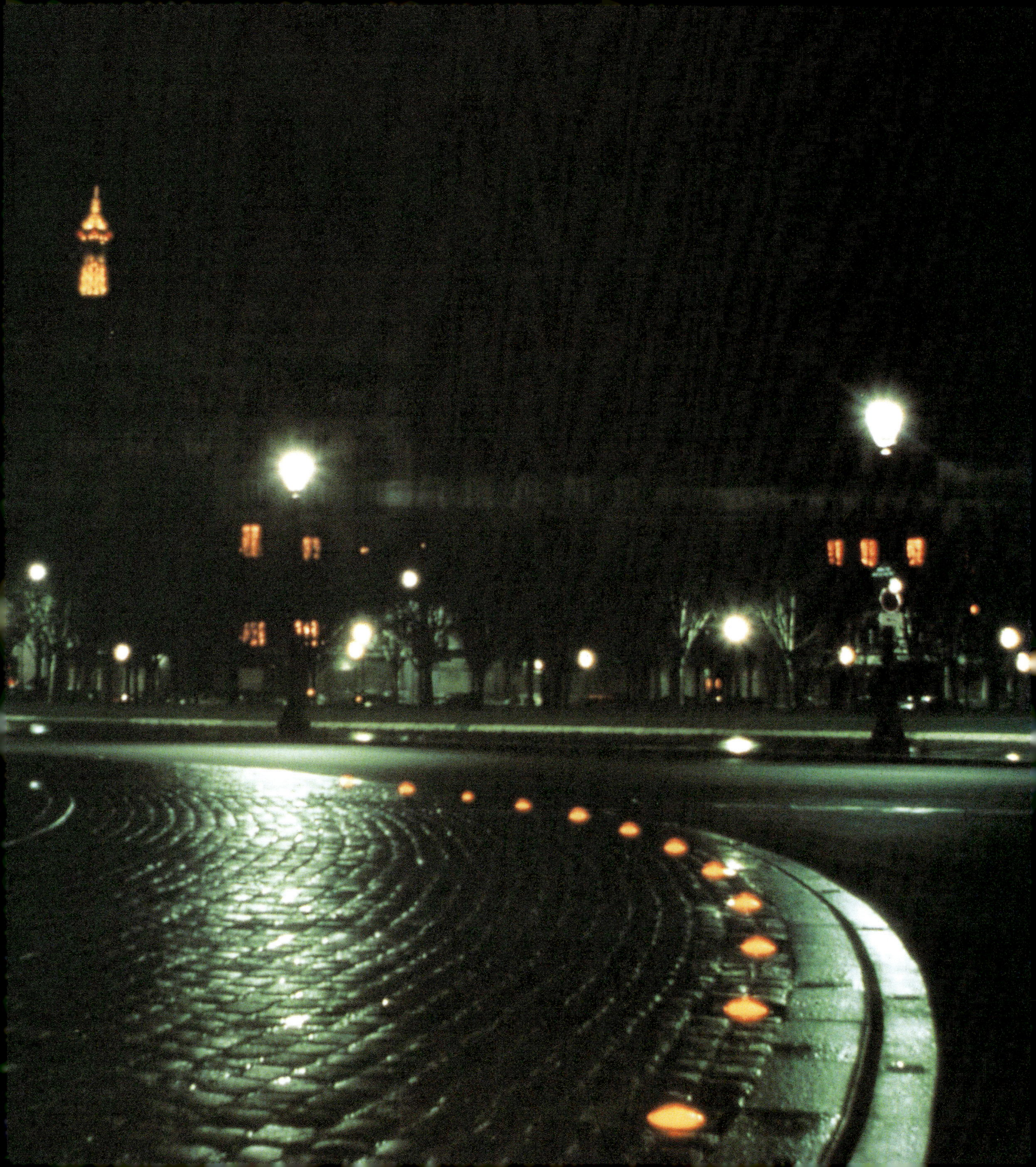

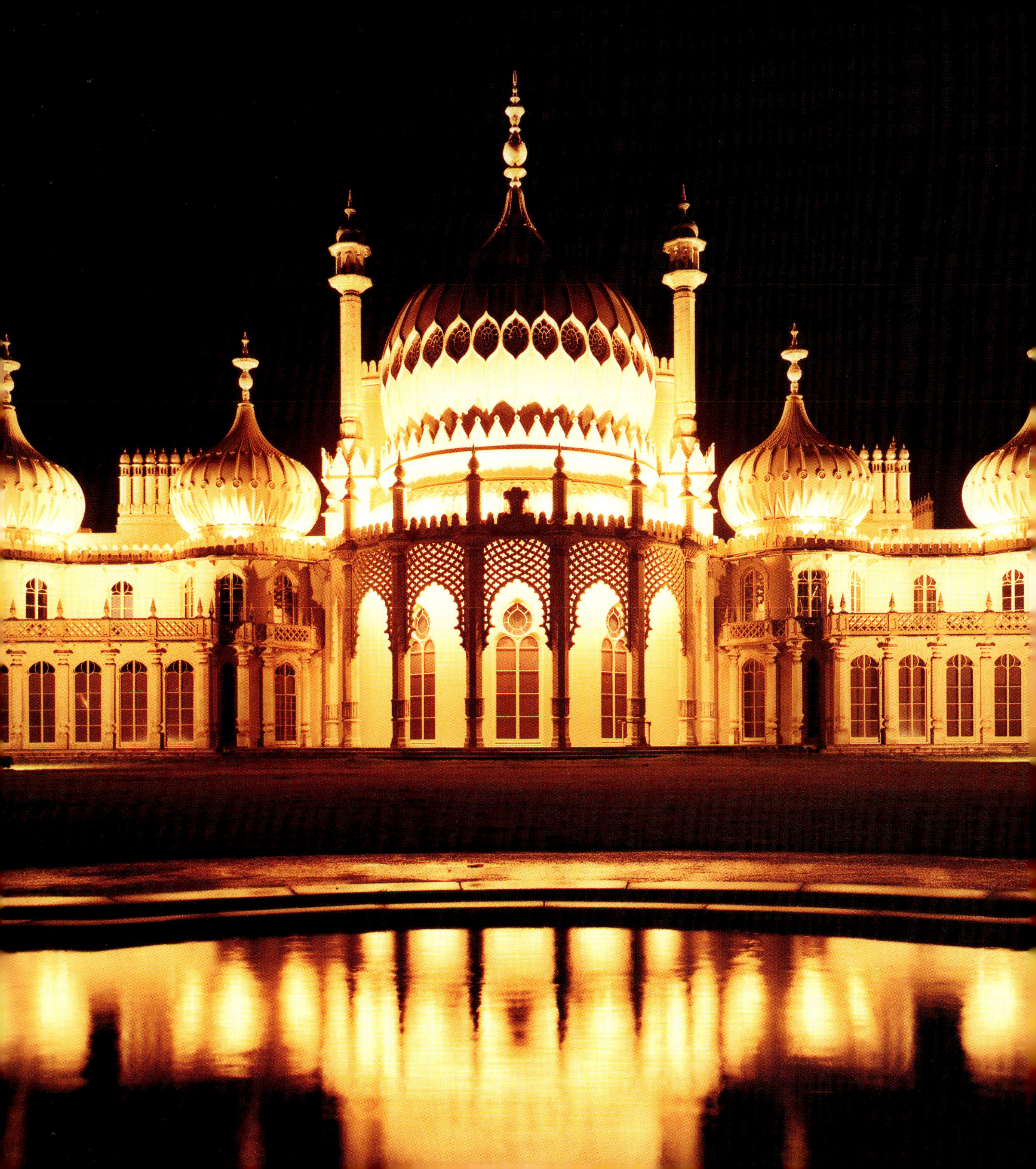

Figure 5.4
The Royal Pavilion, Brighton. Floodlighting of the east elevation. Exterior illumination which enhances and reveals many of the fine architectural details that are unapparent during daylight. Lighting design by John Bradley. Outright winner Lighting Industry Federation 1995 Design Awards.

Exterior lighting and historic buildings

In most cases, historic buildings will require exterior lighting which enhances the architecture and allows surrounding paths and parks to be integrated as part of an overall experience. This consistency of luminous language should begin at the entrance gates and property perimeter.

Some modern bollards may not be compatible with traditional architecture. In the case of a Georgian house, the exterior lighting should usually be of that period. Designers may sometimes prefer to make reference to the historic aspect of a luminaire, using modern design language. In the past twenty years, a thriving industry has developed to supply traditional pole lanterns to the heritage market. In the late 1960s, little such sensibility existed as architects and politicians tore down old buildings, replacing them with quick-pour concrete boxes. Today, however, clients are perhaps too fixated upon the charms of heritage.

Elegant traditional fixtures are available in a range of wattage equipped with the latest in lamp technology. When illuminating the perimeter of a Tudor building, an unobtrusive contemporary design may be more apt than the standard Victorian 'Jack the Ripper' version. The heritage market goes back about one hundred years, though I have seen some jolly carved oak fittings which appeared to be claiming ancestry with Robin Hood.

Whether buildings are old or new, pertinent lighting is a matter of establishing harmony between the building and the topographical vocabulary of the surroundings. Traditional posts and pendant lanterns are frequently specified for main-road illumination in heritage-conscious towns. They are also popular in pedestrian precincts where traditional shops ply for trade. It is a sign of designers' uncertainty that so many are persuaded by the latest sodium-lamp technology, housed in the cast-iron falderals of a Dickensian gas lamp. High-pressure sodium, metal halide and tungsten halogen are the lamps most usually specified for heritage applications. Their use will depend on building type, location and required lamp hours. For instance, high-pressure sodium may suit the warmth of a weathered city wall, while metal halide would help to delineate the stringent exigencies of Georgian buildings.

Figure 5.5
Landscape with water sculpture.
Photographer Douglas Salin.
Exterior illumination which avoids
glare while fully expressing both
form and atmosphere.
Architectural illumination: Linda
Ferry, I.E.S. A.S.I.D. Affil.

Art and exterior lighting

It is possible to create a garden within a garden, illuminated to form an inner sanctum, or private abstracted bower. Lighting can elucidate, suggesting boundaries, yet preserving an ambient link with the surrounding topography. Developing this concept reveals, for example, the possibility of a Zen-influenced garden within a 'Classical' or 'Informal' setting. Water both reveals and conceals, responds to the elements and reflects objects in changed ways. It is an excellent meditative element. Patterns of smooth pebbles, fragments and natural objects could, if illuminated at night, acquire new meaning.

Such patterns are found at the Kettle's Yard Gallery in Cambridge, England, where the diversity of natural beauty is counterpointed with human-made art in such a way that their interrelation transcends mere itemization. Experiments with void and form acquire a complex totemic strength.

Jim Ede, the mind behind Kettle's Yard, was interested in the transforming qualities of light, and the juxtaposition of natural, found objects with those created by artists. His hope was to stimulate a reappraisal of the way in which we react to our surroundings. His extraordinary vision, playing with the possibility of a 'Celtic Zen', has a textural sensuality which encompasses silence and form. By arranging objects into spatial compositions upon the ground, Ede asks us to interrogate the random chaos of nature. His is a synthesis of the artificial and the natural, which establishes connections between metaphysics and perceived reality, embodying the beginnings of ritual.

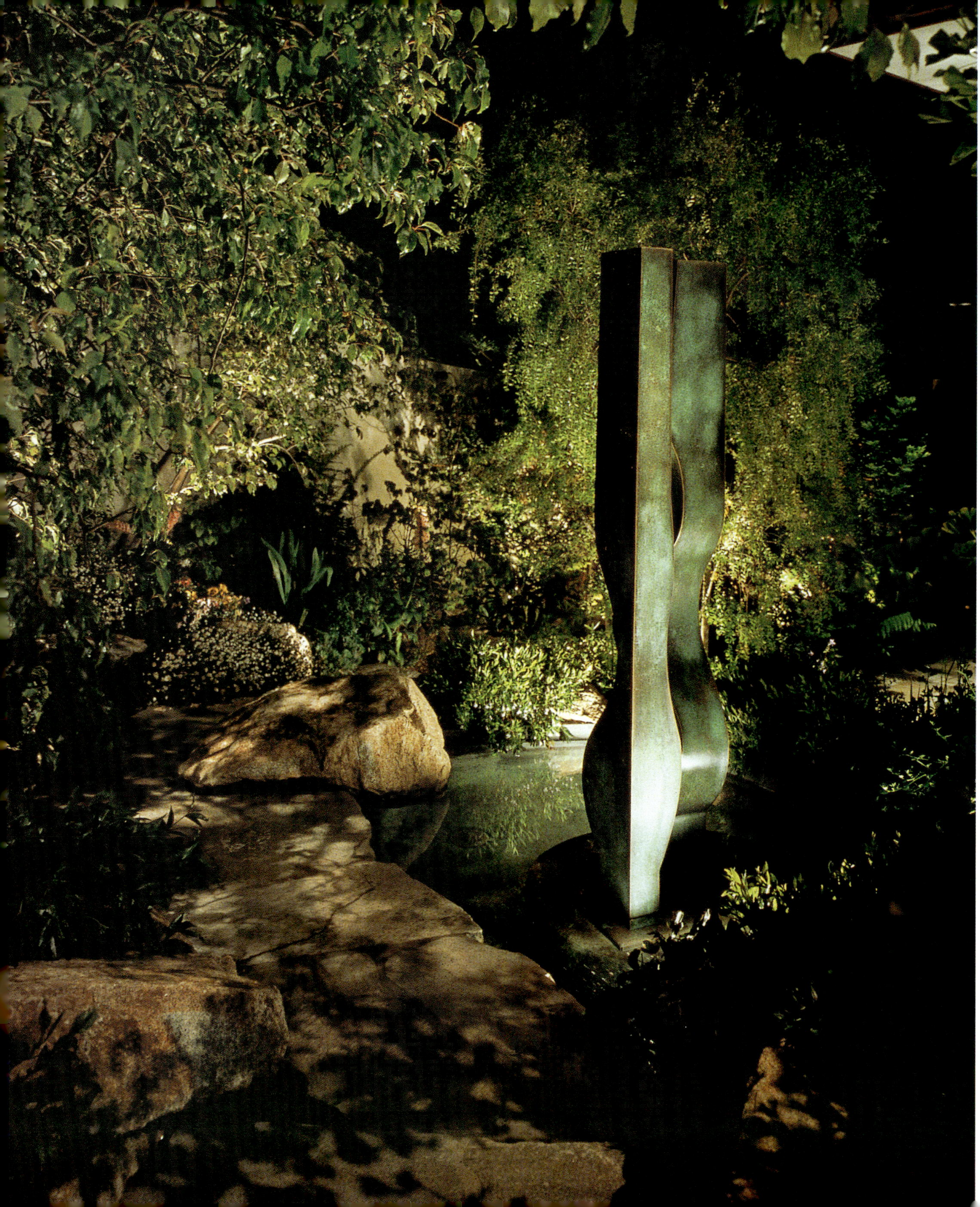

To light such a space requires contemplation and study. It should provoke and intrigue as much by daylight as it does during the hours of darkness. Though only partially revealed by night, contrasting elements will require precise, jewel-like lighting. Fibre optics and low-voltage tungsten halogen are suited to this demanding type of operation. With fibre optics, the light head is compact yet directional and many objects can be illuminated from one remotely located lamp. Fibre-optics' tails are easily concealed beneath the ground. There is no danger of electric shock.

Curvilinear patterns of stones could be traced like vertebrae, in a way that pinpoints each piece as if suspended within the darkness. Looking upwards we may see a spray of leaves, or a vortex of fragrant blossom. Shell fragments shimmer beneath the surface of a pond. When the attention of viewers is focused, objects and the space they inhabit take on new meaning. Shadows can be painted as if with Chinese ink. Narrow-beam tungsten halogen low-voltage sources can direct precise amounts of dimmable light to accent, uplight, silhouette and shadow, creating a fluid, sinuous, nocturnal retreat. With care and consideration, an arena of contemplation can be hidden within the wilder encroaching forms of the surrounding space.

The exterior lighting techniques I have described in this chapter could apply to hotel grounds, recreational facilities, retail landscaping, parks, historic buildings and residential gardens. Any diversion would be one of emphasis and spatial composition. Cultivated environments can be magnificent or intimate, exuberant or pensive. The designer in light must find a way of integrating a carefully calculated permutation of differing aesthetic challenges and with skill and invention set free the particular spirit of that place.

Figure 5.6
Tiled courtyard to front entry door.
Photographer Douglas Salin.
Thomas Lighting. Subtle use of downlighting creates a visual focus within the courtyard.
Architectural illumination: Linda Ferry, I.E.S. A.S,I.D. Affil.

6: Residential

In the family residence, creativity is seldom spent on decorative illumination, except at Christmas, when twinkle light of dissimilar vintages are festooned beneath a beneficent star. There is usually a feeling of sadness when the lights are dismantled in preparation for the short days of January, and a thought that some essential component has once more vanished. It is this positive affirmation which should permeate our living space throughout the year rather than remaining merely an annual harbinger of celebration.

Lighting brings the home alive

Effective interior lighting should not only entertain, for we need humour in our surroundings, but also create a soothing ambience which relieves the tensions of the outside world. I am not advocating Christmas every day, but at times a sense of luminous celebration, in which change and the unexpected can reveal different aspects of the interior, is desirable, though, of course, according to the clients' needs.

    The electric light bulb or GLS lamp has meant that most domestic interiors appear rather static after dark. A suspended pendant is augmented with a table lamp, switched solo for more intimate moments. The set-up in the majority of homes is very similar and very little in the way of creative use of illumination has filtered through to the public.

Early domestic lighting

The history of domestic lighting can be traced from the hollowed-out stone oil lamps of 1500 BC, to rush taper lights dipped in fat, to tallow candles and to the invention of the Argand burner, which greatly improved the luminous output of oil lamps, providing a flame many times brighter than a candle. There was, however, consumer resistance to oil lamps, as it was believed that the fumes were injurious to lungs and eyes. Most large interior spaces continued to be illuminated by candles. In cathedrals and banqueting halls, the combustion of the hundreds of candles required was facilitated by a long thread of gun cotton which ran from one to another, and was twisted around each wick, which had been previously soaked with paraffin. By this method, the candles could all be lit in a short space of time.

Figure 6.1
Contemporary living room. Photographer Douglas Salin. Restrained use of wallwash and downlights balances ambient and accent illumination within a modern interior.

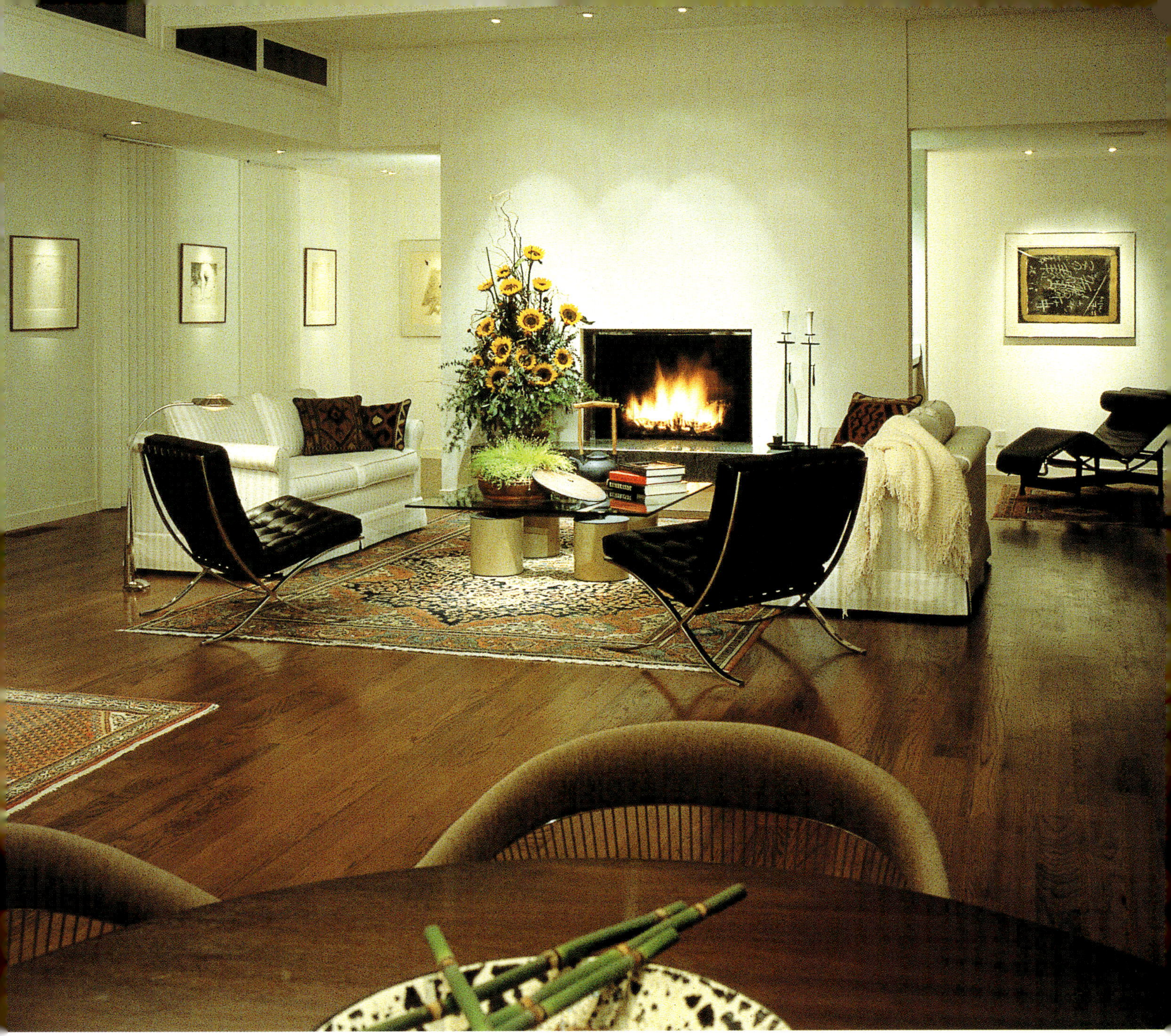

When candles were the only source of illumination, it was as if the void itself had come alive. Even after the introduction of the oil lamp, Ludwig II's passion for the times of Louis XIV caused him to abandon modern lamps in favour of the scintillation of flame, and his gold and azure study in Munich was lit with dozens of candles. In terms of bestowing the quality of intimacy, the candle has proved to be superior to both oil and electricity. Candle-studded chandeliers were often magnificent sculptural creations – their own magical, suspended worlds – concocted from such exotic materials as ruby-glazed glass, silver, gilded bronze, imitation porphry, rock crystal and amber.

Towards the end of the nineteenth century, most houses possessed a paraffin lamp, which required irksome maintenance, but could easily illuminate a small room. Different types of gas burners were also developed and during the 1890s these proved an efficient source of light, though installation costs meant that their use was not universal. During the 1880s, the introduction of electricity, initially in large country houses, meant that gas was gradually phased out.

Electricity arrives

Electric lighting created the demand for a new range of fitments, designed especially for the domestic environment. Louis Comfort Tiffany's (1848–1933) virtuosity with metal and glass delighted the public; he brought a resonant glow to the new-fangled glare of electricity of which many women had complained, saying that electricity's bright light contributed nothing to the delicacy of their complexions. Tiffany lived in Paris and was influenced by his frequent trips to North Africa. His vivid eclectic style, mixing influences of Celtic, Japanese and Moorish cultures, combined to build a bridgehead to Art Nouveau. Sam Bing, Tiffany's associate, first interested the designer in collecting historic decorative glass; it was the same Mr Bing, whose shop in Paris was called 'Art Nouveau', who gave name to and launched the movement.

Art Nouveau caught on like brushfire and, with an easy elegance which adapted well to the new compact light source, turned the shadowy worlds of card tables and cigar smoke into something altogether more visible. In its varied incarnations throughout the world, Art Nouveau facilitated the rapid development of decorative electrical design. Luminaires were chosen not only because they were capable of lighting a space but also because they were pleasing to the eye. Electric lighting made it possible to see more of the domestic interior at night, therefore care was taken in the positioning of luminaires, so that significant features of corridors or halls were revealed.

Residential lighting: designers and clients

There is a difference between the domestic interior which is an integrated part of the architecture, and one which has been designed independently from the protective perimeter. The interior which exists as a different design statement from the shell allows future changes of style and emphasis. The interior which is an essential element of an overall architectural concept, such as Charles Rennie Mackintosh's Hill House or Frank Lloyd Wright's Taliesin West, will, if obliterated, destroy a harmonious cohesion and tacit personal vision.

Frank Lloyd Wright (1867–1959) could create a domestic dwelling which appeared to breathe outward and inward as if the light itself were a source of oxygen. His architectural continuity embraced contrasting textural planes fluidly defined in space. It was his skilled manipulation of natural light which made his domestic buildings such a dynamic spatial experience. Unfortunately, few new homes are designed with the full possibilities of natural light in mind.

Eileen Gray (1878–1977) was an interior designer whose innate sense of void and form was only moderately appreciated during her lifetime. Her work showed a clipped *élan* which formed a counterpoint to much of the machine-like rooms of her peers. Her illumination of lacquer-tiled rooms conjured a vision of interior space contained by its own reflection. The glow and other-worldliness of concealed lighting on lacquer walls created a peculiar, weightless, atmosphere. However, such visions as these are not granted to everyone who wishes to transform an interior space; it is with this understanding that specialists are engaged.

The partnership between painters, sculptors and architects, which was crucial to most of the illustrious interiors of the past, is now less well defined, at times reduced to a snapping menagerie of interior decorators, electrical contractors, architects and project managers. Just as bourgeois intellectuals deplore the aspirations of the gin-and-jeep brigade, so too architects question the fast-track priorities of project managers.

A lighting designer is more usually asked to illuminate the homes of wealthy clients. It is interesting that the scale of these residences poses particular problems. There is a tendency for the space to begin to mimic an up-market hotel, rather than a home. The more money that is spent, the more the subtle overlays of beige and warm white conspire with concealed lighting to mask the personality of the owner. In some cases, this is no doubt desirable. The owner may want a style which appears sophisticated, but is unsure of their own particular needs. An interview with the client is required, rather than with their assistant or project manager.

Interior illumination should complement and inspire. Whatever the parameters of profession and purchasing power, a stranger entering the space should be aware of the personality the client wishes to project: what might be appropriate for a media executive will probably appall an elderly couple with miniature dogs. Even if the budget is generous, it is not always necessary to specify state-of-the-art technology. It is important to remember that the residence is for those who inhabit it and is not a photo-opportunity for magazine-hungry designers. However, risks are acceptable and home owners can sometimes be persuaded to try a lighting approach that they might have imagined was more suitable for people other than themselves.

Figure 6.2
Residential entry bridge with ridge long skylight. Photographer Gil Edelstein. Thomas Lighting. Concealed lighting leads the eye towards the end of the corridor. Care was taken that lamp reflections did not appear on the overhead skylight. Architectural illumination: Linda Ferry I.E.S. A.S.I.D. Affil.

Entry and atmosphere

As with hotels, the manner of entrance illumination will hint at what lies within. The entrance does not necessarily have to be warm and welcoming. The client may not wish their building to be perceived in that kind of way. Doorway lighting could reveal a cool, geometric, precision suited to contemporary stone and steel. However, most people do prefer warm sources which suggest a friendly reception. This can be achieved with either tungsten or high-pressure sodium sources, used to wash the door and an area of surrounding masonry.

In a conspicuously lofty hall it may be appropriate to lower the visual emphasis by means of wall-mounted uplights which partially suggest the void above, but do not fully expose it. Whether the walls are covered in plush damask or the marauding dogfights of avant-garde graffiti, illumination should help maintain a sense of human proportion, so that guests immediately feel at ease. Concealed lighting can bring a glow to the ceiling so that the height is subtly established, rather than starkly revealed. Recessed fluorescent strip, tungsten mini-strip, cold cathode or open-side fibre optic could be used for this situation, depending on the ceiling area to be illuminated. Wall-mounted uplighters can be applied in a similar way in a medium-size hall, though ceiling soffit lighting may not be required.

In the hallway, there should only be one or two accent-lit objects; these will leave a finer impression on the mind than half a dozen craftily illuminated artefacts. Whether the art selected is a goat's head sunk in formaldehyde or an oil painting of Cupid and Psyche will depend on the client's preference. People can feel attracted to the most peculiar objects. The entrance hall gently expands upon the visual statement pronounced at the front door. The two initiate an intricate courtship which will weave, with some silences, throughout the house.

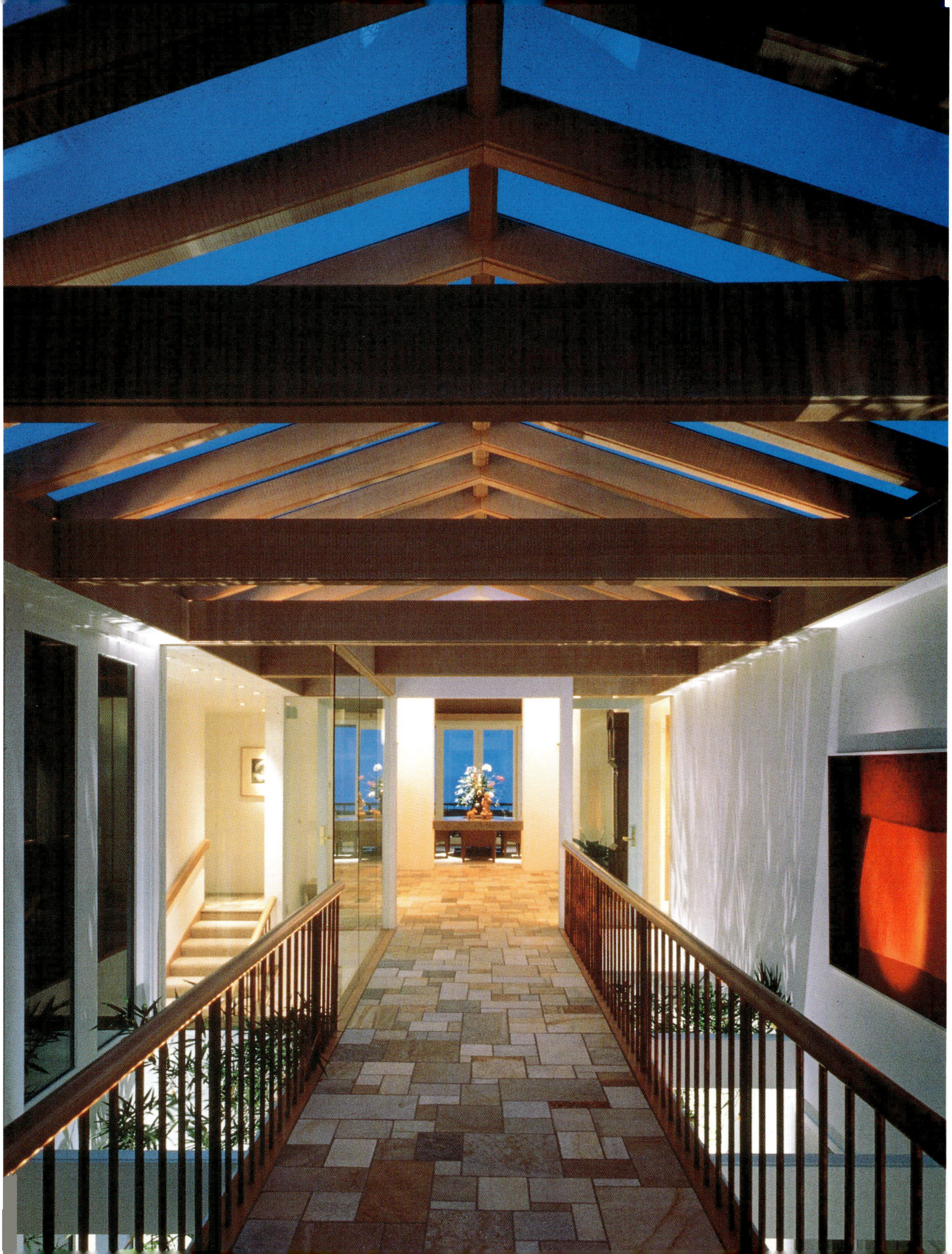

Lighting design in the domestic environment is often a matter of enabling the client to express themselves cogently. There are occasions when the owner wishes a magazine concept grafted on to their home, but even then it is possible to persuade them that perhaps something more personal might be achieved, albeit within parallel parameters.

Light and warmth are central to our most basic needs. This attraction to elemental energy ensures that the primary visual focus in most living rooms will be the fireplace. Those who prefer 'cool' illumination and no fire are in a minority. On a cold evening in a room with a crackling log fire, it may sometimes appear as if the longships of Thor had been consigned to the flames. Firelight and a few scattered candles are often sufficient to illuminate a substantial room. In summer, an attractive fireplace will act as a conversational gathering point, whether the fire is active or not. In a sense, our primal instincts have not changed since the cave: the desire to sit around and tell our stories remains a deep-rooted human need.

Demands of comfort and function must be balanced with lighting flexible enough to handle a wide range of social requirements. Exclusive use of uplighters or downlighters over seating areas should be avoided. Ambient light will help reduce facial shadowing and encourage conversation. Table lamps can provide 'fill' lighting, which softens shadows and reduces surface glare. The creation of a relaxed living space can be difficult, if there is excess accent illumination.

To integrate task lighting with ambient and accent, we should take into account the fact that the fitting will be situated prominently at a table or chair and should therefore be in keeping with the rest of the furniture. For example, a gunmetal grey skeletal task lamp can be tricky to lose among chintzes and chiming carriage clocks. Luckily, task lamps, like other luminaires, are available in a wide range of style and can usually be integrated with most types of interiors.

Task lamps are generally brighter than accent lights, but not to such an extent that they disturb the overall luminous equation. A task lamp will give a directional beam, which is easily adjusted to facilitate activities such as reading or sewing. It can be supplied with a dimmer so that the light level is controllable.

On an executive desk, a task lamp may well be a statement that implies precision and technology, serving as an antennae-like extension of functional business furniture. In the traditional interior, the task lamp will generally be constructed of brass with perhaps a green glass shade. This would be suitable for Victorian interiors. Going further back to a Louis XIV-style interior can present difficulties. A shaded table lamp is the usual solution.

It is important that a task lamp can be easily adjusted and that the beam remains in position. They can look splendid in the lighting emporium, but after a few months' daily use, they often prefer to adjust themselves. The overall standard of construction should be examined before purchase.

The use of electric candles in the living space is very much a question of preference. Even when attached to the most attractive Baroque table lamp, they can appear odd and theatrical during daylight. If the interior is an amalgamation of neo-historicism, the client may insist that electric candles are exactly what is required. At night, when carefully positioned, electric candles lend atmosphere to opulent retrospective surroundings. A compromise would be to place them away from windows during daylight. The artificial candle is specified because it establishes a visual empathy with the king-sized multicultural furniture specially constructed for large residential living spaces. A coffee table of Napoleonic aspirations may have sculptured Egyptian legs, gold Arabic fretwork and a Renaissance marble top. The general feel of the piece is historic, therefore an elaborate electric candle-fitting would seem to stun three birds with one stone.

Other decorative luminaires are wall-mounted translucent uplighters formed from perforated metal or opaline glass. Perforated metal wall lights project an attractive shadow, which dissolves the solidity of the wall and reduces the box-like sense of enclosure. With dimmer control the glow can be adjusted to create changes of mood. Luminaire position should be carefully considered in relation to where main groupings of people will occur.

Venetian glassmasters produce a range of contemporary hand-blown decorative fixtures which enhance both traditional and modern interiors. These take the form of table lights, wall sconces, suspended fittings and avant-garde chandeliers, all available in beautiful colours.

Workmanship of this quality is expensive but exudes such tactile elegance that one piece would be adequate in providing an intriguing focal point within a room. When ordering from specialist European glass studios, it is advisable to have someone with you who can speak the language of the country. These luminaires are often extremely fragile and are easily broken if sent by mail. There is nothing more frustrating than waiting five weeks for a decorative light and then discovering that it has been cracked in transit. It is safer if you can find an agent who stocks the item, though smaller studios do not have an extensive dealer network. You should check that all merchandise will be fully insured while in transit. This will limit the frustration caused by damage.

Accent lighting

Accent lighting should accentuate without obliterating the relationship between decorative and primary illumination. There is a tendency to overstress accent lighting. It is not necessarily the highest note in an aria, but the second violin accompaniment. The highest note might be experienced by looking directly into the light source itself. There can be a confusing similarity between accent and spotlighting. In most cases, a spotlight on stage draws undivided attention to the singer, at the expense of the surrounding cast. Accent lighting allows the complete picture to be perceived intact, so that forms are viewed in relation to each other.

Accent lighting does not have to be confined to objects, but can mean the delineation of cornices, pillars, archways, bay windows and textured surfaces. The luminaire designed for this task is the dimmable, low-voltage, recessed, adjustable, downlight. Track-mounted fittings, fibre-optic and floor-recessed mini-lights can also be utilized for accent lighting, depending on the space and character of interior to be revealed. For instance, a traditional space would not usually be lit with metal halide recessed downlighters. There would be a pendant luminaire and table or wall sources, with recessed downlights used to enhance illumination levels at axis points, such as archways or floor changes. The traditional interior responds well to a mix of different illumination sources, so long as they are orchestrated to produce visual harmony.

Lighting art objects

Owners sometimes desire an abundance of art to be highlighted. It is much more interesting if only two or three objects are selected for special attention. They should be positioned so that they are noticed in a natural sequential order and not because they lie pinned beneath an interrogator-type spotlight. It is advisable to specify dimming control on all accent lighting, so that levels can adjusted against the task and ambient fixtures. It may be that the object you wish to accentuate will be changed occasionally, so a different level of light is required.

One method of lighting art objects, especially if they are in an alcove or on recessed shelves, is by using low-voltage tungsten miniature lamps connected in a strip, and recessed beneath straight or curved surfaces. They are excellent for producing a glow within restricted spaces. Open-side fibre optic can also be used for this purpose. There is a possibility with a larger shelving unit that too many objects might be illuminated, creating a distracting, department-store, effect. This can be avoided if the collection of books and objects is apparently random, giving the appearance of a slightly chaotic working study. If book titles require to be easily read and the bindings are not especially attractive, they should be lit with ambient light rather than being highlighted.

Lighting specific residential spaces

The living room

The living room in the family home is a multi-purpose space in which reading, television viewing, laptop lunch, board games and children's play may all occur within twenty four hours. With children on board, the living room can rapidly resemble a post-hurricane Disneyland. If there are teenagers in the family, there will be a need to diminish the visual impact of youthful debris, achieved by a rapid tidy-up, then dimming all but essential task lights. A professional couple may keep a harmonious whimsical room, uncluttered by the strewn appurtenances of youth. In both situations, there will be a need for a variety of lighting options which respond directly to particular needs.

In the living room, the lamps most frequently specified are tungsten halogen and incandescent GLS. People wishing to transform a conventional living space with central pendant and shaded table lamps might consider some basic tactics. The first would be to lower

Figure 6.3
White living room overlooking the ocean. Photographer Douglas Salin. Thomas Lighting. Use of chandelier and wall sconce illumination to create a relaxed residential environment. Architectural illumination: Linda Ferry I.E.S. A.S.I.D.

the room's spatial sightlines by dispensing with the central pendant and creating diagonal washes of light with table and floor lamps. This helps to soften the rectangular uniformity of most domestic interiors. Another possibility is to source a central pendant which contributes effectively to the character of the room. This in conjunction with dimmable wall sconces will create a more flexible luminous solution. It is seldom that Chinese paper lanterns add anything to a room, and in most cases are chosen because they can be relied upon to detract the least.

Accent lighting, perhaps in the form of a low-voltage recessed downlighter with tungsten halogen lamp, might be used to draw attention to a painting or some architectural feature such as a fine fireplace. If a painting is to be highlighted, it should be a work of some merit and worth looking at, so that, like a subtle cypher, it echoes the taste and preoccupations of the owner. It is a care for the meanings revealed by illumination which establishes a sense of authenticity, and helps to connect voids with artefacts. Good lighting is not always a matter of spending large sums of money, but of creating an affirmative environment.

The dining room

The dining room sometimes doubles as library, writing workroom, or informal picture gallery. In smaller homes, the dining area is incorporated within the living room, or in a through-space or hallway. Illumination should enable guests to perceive each other and the food before them in an appealing and appetizing manner. With this in mind, a cut-glass chandelier is often suspended above the table. This will give a soft, diffuse, yet vibrant, illumination without shadowing, which is good for conversational situations.

When guests are present the table may be decorated with cut glass, candles and flowers, creating a charming *mise en scène* within the confines of the house. As the dining room is not in continual use and the chandelier's electric candles are positioned above eye level, they will appear less artificial than when placed on a table in daylight. In most medium-sized rooms, the chandelier should not be suspended any higher than six feet above the floor. This helps to suggest intimacy and enclosure. Taper candles can be placed on the table to add real flame to the chandelier's sparkling aura. Ambient, dimmable, wall sconces will provide 'fill' lighting and create a sense of soft perimeter enclosure. Directional downlights should not be sited above the diners' heads, unless there is sufficient background illumination to cancel the effect of facial shadows. In situations where a chandelier is inappropriate, downlights can light the table area only. Candles can then be lit to counteract any peripheral shadowing that might occur on guests' features. Some modern chandeliers have an uplight incorporated into the cut-glass corona, thus adding to possible combinations of reflective ceiling light. Illumination output should not be too high, as a chandelier is meant to glow and scintillate. However, it should appear as if the main source of light is the chandelier.

Accent lighting on art objects should be controlled so as not to distract diners. It can be irritating during conversation to be aware of reflected light from ceramics or picture glass. Whether you are dining beneath a modernistic chandelier with Medusan extensions of twisted glass, or at either side of a three-branch candlestick from the local trinket shop, the nub of the experience should be one of harmony and unity. Like moths to the flame or hyenas to the fire, we are drawn by the scintillation of light, whether it be the sun's last rays on a fine salmon beat, or a weightless trance of Bohemian glass.

In residences where a chandelier is inappropriate, a dimmable pendant in alliance with table-top candles can offer an effective compromise. Track-mounted spotlights should be discouraged as they are a distraction, hanging like so many metal unmentionables, pegged to a vitrified clothes-line. The light fitting should focus attention on the table top and establish a tactile empathy with the objects arranged there.

The crystal of a chandelier or contemporary layered glass pendant shares the same reflective language as wine goblets, decanters and silver cutlery. In a sense we are creating a round cone of unity, within which the diners sit protected. The informal proprieties of modern dining mean that an atmosphere of humour and relaxation is much preferred to the rigorous formalities of yesteryear. The usual intention is to encourage conviviality, while impressing without ostentation. The table illumination is primary, with other select objects lit in such a way that they are quietly registered during the course of the meal.

The kitchen

The powerhouse that drives the dining room is undoubtedly the kitchen, and the culinary delights that descend on to the table will in part be due to the kitchen's operational efficiency. The kitchen should function not only as a place in which to cook but as a place of retreat, where coffee can be brewed and newspapers perused in peace. To design a kitchen merely as a fully fitted locker room is to miss the element of community essential to its creative existence. Illumination requirements should encompass activities which may not be directly concerned with food preparation. This should be discussed at length with the client so that specific needs can be targeted.

Tracked spotlights should be used with discretion, as they will cause difficulty with shadowing unless carefully positioned and fitted with adequate anti-glare baffles. Shadowing on preparation surfaces should be avoided. There can also be glare from reflective worktops and this can occur whether the material is light or dark in tone. Lamps utilized should not be too cold in colour appearance. This would separate the kitchen from the rest of the residence. A luminous continuity should exist throughout the rooms. Some clients may wish to contrast a warm or cool appearance, as a visual counterpoint within the building. This technique is more effective in through-spaces or hallways, and must be carefully orchestrated, with either warm or cool predominating.

There may be the temptation to give the kitchen the look of a culinary laboratory, but food prepared in this space should have a similar appearance when taken to the dining room. The first impression of food is crucial to its initial reception. This is not to say that you cannot have a crisply illuminated kitchen, but an excessive disparity in the colour-rendition properties of lamps used in kitchen and dining area should be avoided.

Lamps suitable for use in kitchens include low-voltage tungsten halogen, incandescent GLS, mini-strip xenon and fluorescents suitable for use in culinary preparation. The main consideration should be lack of glare, minimal shadowing and a sense of visual connection with the rest of the residence. Diffuse ambient light can be provided by wall sconces, minifluorescents, and miniature lamp strip-lighting, recessed beneath cabinets or work surfaces. There should be dimming control on ambient and direct sources. Direct lighting usually takes the form of recessed adjustable downlighters, or recessed fluorescents with prismatic filters.

If task lamps are specified, they should be positioned so that they do not obstruct preparation surfaces. In the executive apartment, the kitchen may well be an enclave of parallel, machined-metal, surfaces, where the making of a peanut butter sandwich might well be construed as an act of desecration. In some situations, there may be a design decision to discourage activities other than the efficient preparation of food. By contrast, the domestic family kitchen is usually a colour wheel of folk art piracy, full of incident and fun, barely managing to cling to the coat tails of tidiness. This kind of environment will require warmer illumination which considers the needs of partition for a range of different activities as diverse as craftwork and bicycle repair.

The bedroom

Lighting in the bedroom will also depend on the type of activities which take place there. For most people, their time in the bedroom is spent unconscious, while others telephone, eat meals, watch television, read newspapers and conduct business. The lighting designer must ascertain which are applicable.

The lighting scenario will depend on whether the bedroom figures as an extension of sleep, or whether it is a springboard to the waiting day. In most homes, the domestic bedroom functions as a medium-range hotel bedroom, and illumination requirements will be similar. There will be a variety of switching options available from the bedhead, including dimming control and en suite bathroom illumination. It will be quirky idiosyncrasies of ambient and accent lighting which will effectively distance the space from the mindless pastiche of most drive-in highway hotels. The bedroom is a place where we are at our most vulnerable, unconscious and in dreams. It is a private space reserved for the workings of the subconscious, but this aspect is usually denied. It is rarely that a connection with fantasy is explored. More often we pursue the bland functionalism of a catalogue, complete with matching bedroom suite. Most clients are fearful of venturing into unknown territory, which is a pity as the illumination prospects could be intriguing.

For a couple with young children, the bedroom is a final refuge, which doubles as an early morning meeting place and child–adult dressing area. In this situation that delightful hand-blown Venetian piece is best forgotten, as it will only pose a safety hazard. Table lamps should be solid enough to withstand tiny hands. Electricity does not discriminate. Care should be taken that there are no looped electrical flexes which children could trip over. Wall-mounted uplights should be high enough, so that they cannot be grasped by a child on a chair. Lamp sources specified in the bedroom are usually of a warm colour appearance and include incandescent GLS, low-voltage tungsten halogen, and warm-bias fluorescents, often of the compact variety. Lamp sources will either be shaded or fitted with diffusers or baffles for soft indirect light.

Figure 6.4
Landscape with pond and stones. Photographer Douglas Salin. Fixtures are hidden from view and focused at a variety of angles to illuminate the scene for interest, balance and depth. Architectural illumination: Linda Ferry I.E.S. A.S.I.D.

The bathroom

The bathroom is the space where you can ascertain whether the owner is committed to design. In this space, character and quality can vanish, replaced by shiny moulded plastic and bathmats embossed with sailboats. A bathroom should be as creatively illuminated as the hall or living area. It is the space we first encounter upon rising and it may well affect our mood for the rest of the day; a bathroom should cosset and prepare us for the outside world. Lighting and mirrors should be positioned so that skin tones appear healthy. Our facial appearance in the mirror can be drastically affected by angle and quality of light. A north sky light will cause shadowing and pale skin tone. In some situations the person will actually seem to be unwell. In a bathroom that receives direct sunlight, semi-translucent shades will absorb excess glare. The lighting designer should check the bathroom's orientation. The duration and angle of daylight should be considered.

A diffuse light source on either side of the mirror will help to cancel facial shadowing. These can be concealed mini-fluorescent strip or wall-mounted translucent sconces of alabaster or opaled glass. A light should not be fitted directly above the mirror, as this will cause shadowing.

For illumination of the bath area, recessed adjustable downlights – angled to create reflected cross-lighting in conjunction with dimmed wall sconces which utilize the ceiling as partial reflector – will help establish a luminous ambience suitable for contemplative bathing. If required, compact fluorescents could be fitted in wall lights, if their colour rendition was suitable for skin tones. All luminaires should satisfy electrical standards for use in interior 'wet areas'. As these requirements will vary from country to country, it is important to check with the manufacturer that the fitting has passed the requisite tests for use in bathrooms. Dimming control on both ambient and direct sources will allow the creation of different luminous environments. These may range from a cool pristine exactitude, to a shell-pink effervescence synonymous with lascivious sojourns in deep sunken baths.

Plants, paintings and prints can bring a zestful personal touch to what can be a rather characterless space. In a way, we are trying to introduce texture, interest and colour, perhaps in a less formal manner than experienced elsewhere. We do not have to worry here with the pusillanimous preoccupations of society's art slaves, where humour and humanity is abandoned in favour of what is fashionable. The zany diversions of vaudeville or photogravures of Palladian villas may well suffice. Whatever the client's inclination, the result should entertain. For too long, bathrooms have been plastic havens, articulated with chrome or gilded caravan fittings.

Further lighting possibilities include fibre optics as a concealed linear source beneath shelves or fittings, or as a sparkle point above multi-coloured bath bottles. Depending on the size of the bathroom, gobos could project patterns in conjunction with colour-filtered recessed adjustable downlights. There are many luminous possibilities, without sinking into a sub-oceanic discoland. It is a matter of planning and having a consistent vision of what is required. The bathroom should be as much a part of the house in design terms as the kitchen or dining room. Attention to the quality of lighting is a positive step towards achieving this aim.

The house should be a refuge and source of inspiration, nurturing and preparing us for the world outside. It is a reflection of self and our ongoing state of mind. The house will reveal our idiosyncrasies, preoccupations, weaknesses and strengths. It is the place where we felt most at home as a child, and also where most of us would hope to die. It does not have to be large – it could be a three-room apartment, a caravan or even a boat. What matters is that in some way it talks to us about our life. These hints and nuances separate the house from the most sumptuous of hotel rooms: though the house may empty on our death, its presence will remain recorded in many minds.

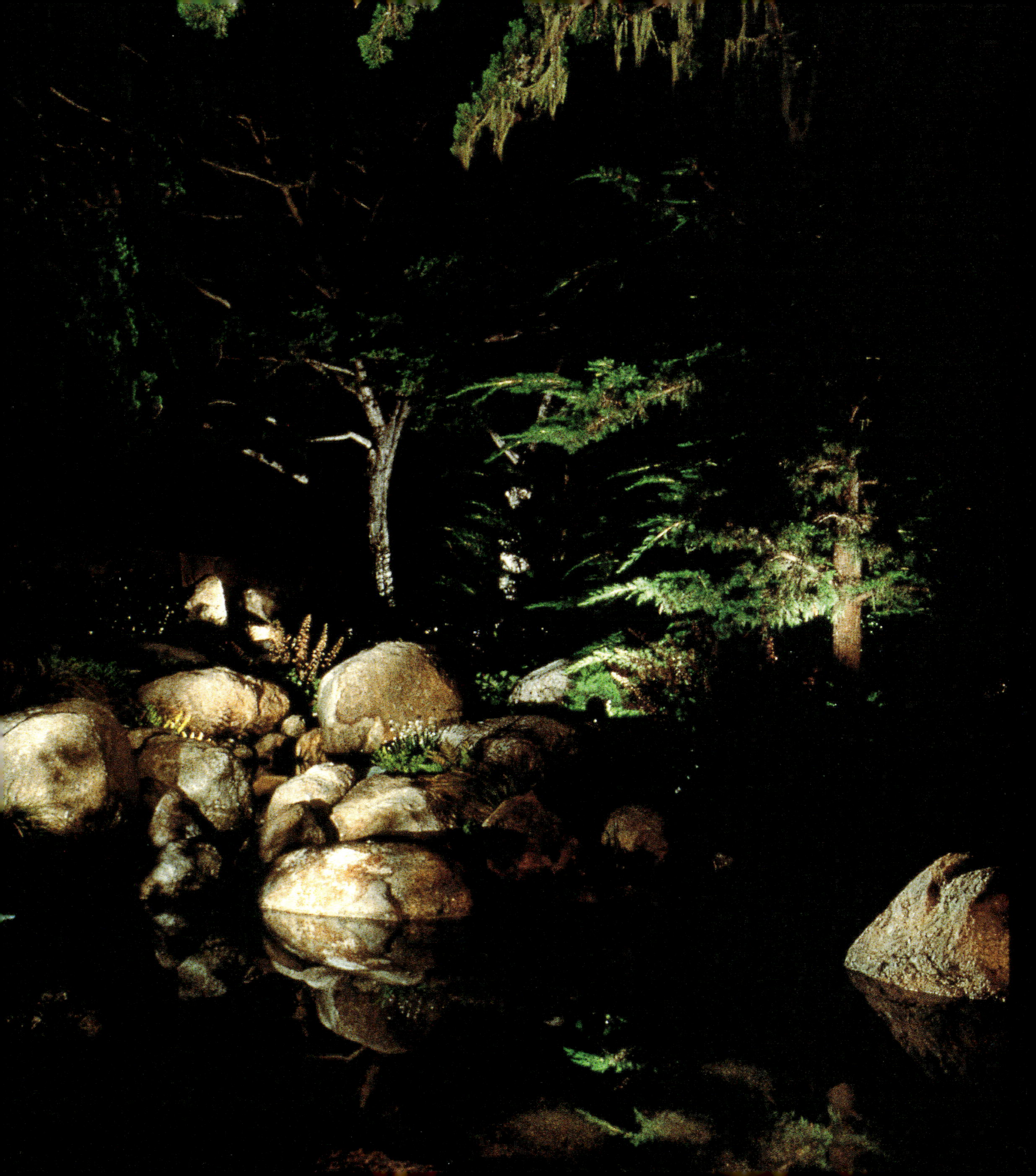

7: Past and Future Predictions

Light, art and architecture

It was not until the end of the nineteenth century in Germany that a philosophy of space dynamics began to examine the theoretical relationship of form and void. Planes within space revealed by light articulate the essence of architectural accomplishment. Light provides connection between outer shell and interior. At the same time as philosophers developed this esoteric cabbala of void, electricity was introduced to domestic and commercial premises.

Theories of beauty and conceptual space were postulated which rarely included light, except when encompassing theological dogma. The great Gothic churches, with their beguiling harmonies and phalanxes of glowing glass were deeply emotional buildings, rather than speculative intellectual exercises. Church authorities were not known for encouraging radical thought. By contrast, German intellectuals appeared to sense that there were great changes coming, which would alter our perception of the universe. The publication of Einstein's theories regarding space-time provided potent primary cyphers for the experimental work of German Expressionist architects and artists. The ideal of the factory as shrine to the machine, devoid of decorative anachronisms, developed into a belief that architecture could provide the stepping-stones towards an improved society.

Architects such as Walter Gropius (1883–1969) endeavoured to integrate painting, music and dance within an overall philosophy of architecture. Gropius and Mies van der Rohe (1886–1969) used light and space in a way which encapsulated mercurial brilliance within the strictures of a rigid glass and steel grid. The new discoveries of Cubism and the desire to express a fourth dimension influenced the volumetric geometry of German Expressionist architects. As in the canvases of Mondrian, the energy of these interiors slowly develops from the core to the outer limits of the envelope, and then perhaps extends into space. It was an unprecedented assertion of design which disregarded the presumptions of architecture contained within an ornate frame.

Developments in luminaire capabilities, allied with the iridescent traceries of Art Nouveau, helped to free imaginations from the predictable introspective symbolism of the past. The influence of Expressionist and Cubist painters, such as Braque, Kandinsky, Klee and Picasso encouraged the creative dissection of electrically lit interiors. It would take some twenty years before functionalism became an excuse for uninteresting architecture. The exchange of concepts between the arts brought a dynamic use of mass and space, which at times achieved a Jazz-Age Dionysian freedom.

Figure 7.1
The Kunstbau exhibition hall of the Villa Lenbach in Munich. Architect Uwe Keisser, Munich. Lighting design by Lichtdesign Ingenieurgesellschaft mbH, Cologne, Germany. Neon artist Dan Flavin crates an energized cyberspace in what was once a subway hall. Erco Lighting Limited.

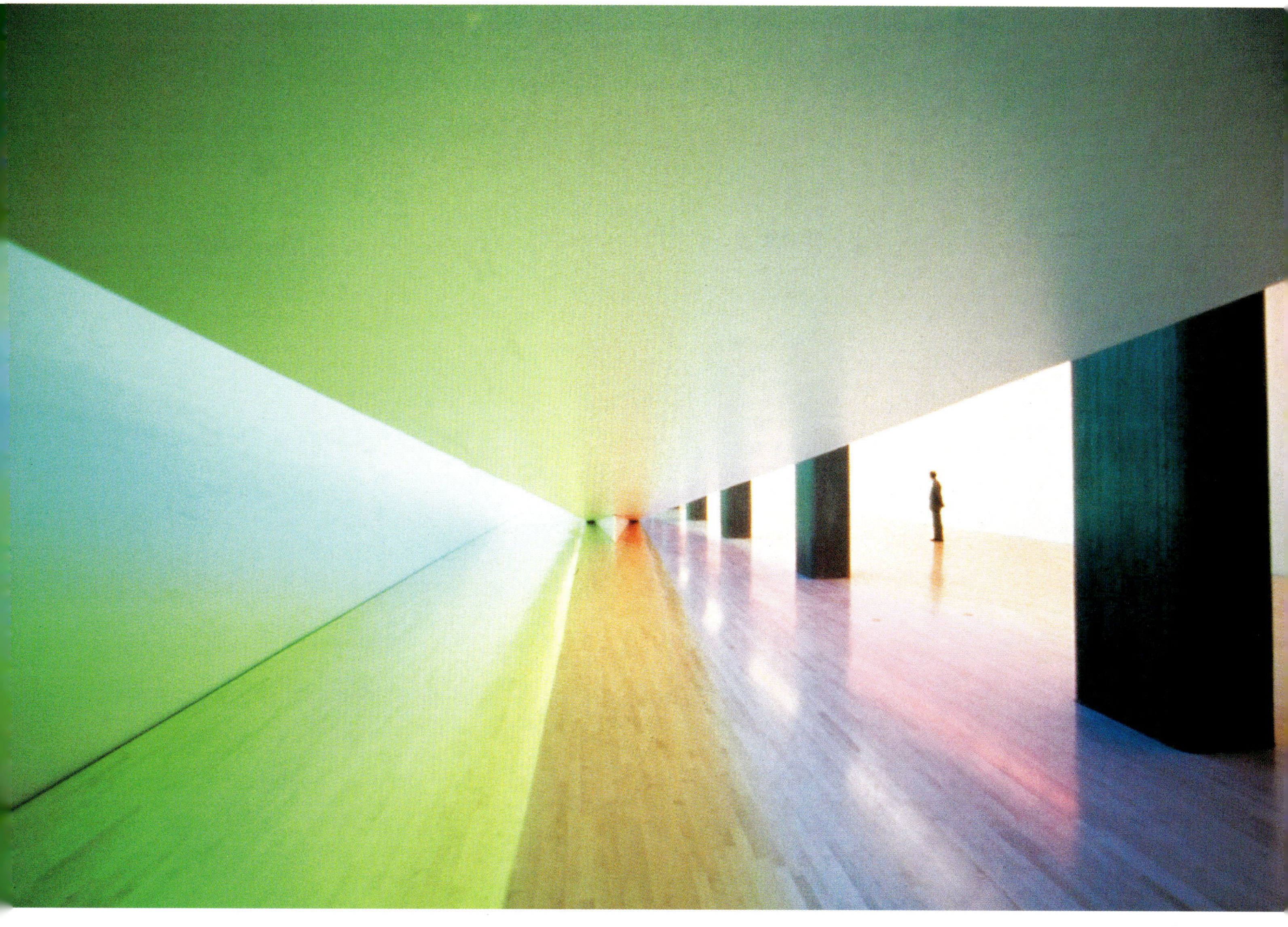

Through the increased use of glass, the interplay of curved and angular planes brought a sense of weightlessness to complex tonalities of light and shade. In the past, architecture might have been described as the articulation of enclosed concave space, which on occasion became a bucolic bonanza of ornate surface decoration. The spirit of the 1920s would not be so readily harnessed to flights of plaster 'putti'. Mies van der Rohe believed that a building could transcend its massed state of steel and concrete and enter into a realm of infinite space, realized as art. The underpinning to this vision was a synthesis of space, form and linear elements, transfigured through the new miracle of visibility. In essence, glass introduced an oriental dimension, which cut through to the westerner's inherent fear of void and weightlessness.

The industrial and cultural upheaval at the beginning of the twentieth century caused seismic reverberations through all sections of society. Though a street organ grinder might not be affected by Einstein's paper on 'Electrodynamics and Moving Bodies', he would be aware of Cubist posters, jazz, photography, the telephone, motor cars and aircraft. This extraordinary burgeoning of artistic and technical advances, coupled with the novelty of electricity, signalled a speed-driven world which was already on the move.

Given the intellectual theories of space developed by German architects, it is interesting that the most skilled use of natural illumination was achieved by an architect who was influenced by the organic interrelation of form. Neither mass nor void was an exclusive design priority for Frank Lloyd Wright, yet his understanding of light revealed a depth of meaning unique in the annals of contemporary architecture. Considering the preoccupations of the time with cuboid structures, the plasticity of his vision seems all the more remarkable.

Refining lighting technology

Today, for artificial or natural light to be managed in truly novel ways would probably require a transformation in the basic experience of interior space. However, within the existing parameters of architectural design, we can expect steady progress in the refining of lighting technology. The flexibility of intelligent interactive illumination, which reduces energy costs and meets specific user needs, will increase in levels of sophistication, and perhaps bring to a close the dichotomy between light quality and lamp efficiency. Spin-offs from electronic research will feed into commercial lighting. A cost-effective light source may be discovered which heralds the demise of the domestic GLS lamp.

Indications are that solar power will supplement electric systems, especially in larger government and commercial projects where clients wish-to be perceived as environmentally friendly. The transference of solar technology to the domestic sector is usually slow unless subsidies exist to encourage it. Illumination technology is not as readily absorbed by the domestic market as are advances in audio and communications. Quality of lighting is not generally perceived as a priority compared with in-home audio and video entertainment. While consumers skip track and channel with remote control, and think nothing of it, they usually do this within a room lit by the time-honoured GLS lamp. If the domestic market was ready for a handset-operated light-scene controller, we would undoubtedly have one. It may require the development of a compact multi-directional automated light system which bypasses conventional wired routes, before the public are sufficiently fascinated to invest in the latest technology.

The stacked portable audio module quickly made obsolete the multi-wired lateral hi-fi system. The average life of a portable audio module is four to five years, while the old-style lateral system lasted over ten years. It is perhaps worth questioning the so-called long-term quality priorities of commerce. It is clearly not in the interests of lamp manufacturers to develop a domestic product which lasts indefinitely. What I believe we can expect is a lamp with perhaps three times the life of a GLS, which provides good colour rendition, and can be manufactured at a comparable cost.

The idea of light quality corresponds with current health concerns and if a correlation between full spectrum light and improved health could be scientifically established, then the public would respond. The depressive condition called Seasonal Affective Disorder (SAD), caused by lack of sunlight, does undoubtedly exist, but is still perceived by the general population as a grey area akin to ME.

In recent illumination of hotels and restaurants, there has been a move towards recreating an atmosphere similar to the Vienna Secession and Hagenbund. Designers such as Joseph Urban (1872–1933) eschewed the traditional antlered halls of the Austrian aristocracy for a sweeping and sculptural vision, which later translated well to the American stage. This theatrical appreciation of scale and occasion has become a manifestation of recent European design, as if the elegant tonal beauty of black and white films made in the 1930s had been rediscovered as an appropriate vehicle for establishing a sense of time and place.

Hotel and restaurant proprietors realize that style remains while fashions change. However, a concern for refinement does not mean mindless pastiche but rather a development which acknowledges the importance of scale and detail. Perhaps the twentieth century has become over-heated in its need to astonish and might benefit from exploring a basic understanding of line and form. The artist steals, but always with a view to development. At times of crisis, history invariably loops back to re-establish itself upon the foundations of the past. For example, Hotel Moderne is a refined development of Deco.

Projections upon the future of illumination must consider the implications of past prophecy. Writers in the mid-1960s discussing the long-term effect of global communications were if anything somewhat conservative in their predictions. The advance of many specialized areas of technology has been phenomenal, though few surpass the global communications network. Ironically, space flight engineering has at times almost come to a standstill. Progress within the lighting industry has been steady, though hardly newsworthy.

Lighting and the future of the workplace

The Italian Futurists, writing in 1913, envisaged a world of frantic, chaotic energy, which manifested itself as much in random violence as it did in unbridled technical excess. They wished to make the impossible a reality and in many ways their vision has been realized. Modern technology has brought with it psychological and social problems. An appreciation among business circles of the benefits of appropriate lighting in the pressured work place could ensure long-term damage limitation among both employees and management.

The need for employees to succeed in their work has never been greater. The concept of a job for life is now a thing of the past. The young live with uncertainty as part-time staff replace secure positions. There is a growing awareness of the need for self-maintenance, through healthy diet, exercise and an understanding of the importance of our domestic and commercial environments. The quality of interior space has never been more crucial. It is essential to the way we perceive ourselves, and our employer's relationship with us. Often companies which operate health-care screening pay scant attention to the way light reveals the workplace and surroundings. If the management ignore quality of illumination, beyond providing adequate light to perform tasks, then the long-term outcome is often absenteeism and loss of productivity.

A future role for illumination will not only be energy conservation but the active psychological support of staff. As industries become more specialized, the retraining of replacement employees more expensive, and employees increasingly hired on short-term contracts, the provision of an affirmative working environment will be a vital factor in remaining competitive. Workers who perform high-speed repetitive jobs in characterless surroundings are especially prone to long-term psychological stress. I am not suggesting floral upholsteries or the overstuffed trappings of a suburban home, but rather a compassionate and real awareness of what the workers actually have to do, thus providing a congenial platform which is suitable for work.

The future for lighting design

With indications of a global fracturing of society, not only into strata of haves and have nots, but also into smaller ethnic and cultural demarcations, it seems that advanced illumination technology will not be a feast consumed by all. While processor-driven illumination has integrated smoothly within the burnished mirror-glass techno-rafts of multinational corporations, millions of homes throughout the world have only a tenuous contact with the benefits of electric lighting.

To speculate on the future requires balancing what are often conflicting bodies of evidence. We can underestimate the effect of runaway technology, as much as we do the ability of war and disease to alter the course of history. On examining the past it is apparent that architects and artists often have precognition of what is to come. Many of the paradigms of architectural power sketched by the Italian Futurists depicted the city skies alive with curves and angles, as if subconsciously they saw the myriad frequencies of satellite transmissions. Luigi Russolo's futurist painting, 'The Houses Continue Into The Sky' (1913), is an interesting development of this idea of connection between the habitat and the heavens.

The perspective drawings of Antonio Sant 'Elia were in part predictions of forms which would be built on a reduced scale during the 1920s and 1930s. In Los Angeles, the Academy Theatre, Pan Pacific Auditorium and LA Public Library could all have been developed from Futurist sketches of 1912.

The vital role for lighting designers

There has been a tendency for architects to regard lighting designers as back-street counter-jumpers whose forced entrance to 'Athena Ergene' was effected while the priests of the temple slept. Nothing could be further from the truth, for compared with the pernicious regimes of 'Design and Build', the inclusion of 'lighting design' specialists on projects has established the importance of illumination on a scale previously unknown. Lighting designers understand the possibilities and beauty of space, in a way which is alien to 'Design and Build' project managers. Taking a hard-nosed look at what lighting specialists can achieve in the commercial sector, reveals instant productivity rises of up to 30 per cent, product rejection rate down by 80 per cent, energy savings of 40 per cent or more, plus marked reduction in absenteeism. Admittedly, some of these targets could have been achieved by electrical engineers, but the lighting designer directly addresses quality of environment and the subtle business of creating spaces in which people can work comfortably and effectively. It is a well-calculated but holistic approach.

Though technology mutates inexorably, with a gargantuan momentum of its own, the means by which this can be harnessed to provide beneficial results for people in the home are often slow in realization. The idea of global unity remains a dream. Against a background of cultural and political difference we see increased technological uniformity, not only in matters relating to illumination, but in the universal language of computers and communications systems. The agreement and harmony of our machines stands in contrast to the efforts of peacemakers.

In Japan, a country culturally defined by an obsessive need for social conformity, their architecture and deployment of both natural and artificial light, has never been more quirky and wonderfully individual. In the past ten years, some of the most exciting architecture and lighting has emanated from Japan. There is a sense of style and risk-taking which encompasses a fusion of constructional tradition and an anarchic symbiosis of modern materials. Within a refined architectural geometry which mirrors the building blocks of nature, there is a historic understanding of impermanence. Buildings are not designed to last forever, but to be renewed by a further dynamic vocabulary which speaks of tradition and technology. Architecture in itself is not an art, unless those involved in its creation have the intuition and vision of artists.

The human spirit is forever mobile and it is this restlessness and need to come to terms with chaos which enables progress. Light is a key element in shaping interiors both dynamic and contemplative. Light can charge the working situation with energy and zest, whether that be a downtown beauty parlour or a high-rise office block. The number of employees may be ten or a thousand, but their needs remain the same. There is a habit of thinking that may have lingered on from the days of the Industrial Revolution, that the more employees you put together, the less affected they will be by their surroundings. People do notice the quality of the commercial environment and if a sense of team spirit is to be preserved, then managers would be advised to pay particular attention to illumination. Light can bring unity to areas of complexity while allowing layers of tonal ambience, which helps to dispel the pitfalls of workspace monotony.

Through the many ages, humankind has relied on light for protection from enemies and the growth and gathering of food. Without light there would be no sustenance. Light has been taken for granted, but its removal for more than a week would signal the end of the human race. No matter how much we postulate about the intriguing diversity of space, its understanding would be difficult without light. Our relationship with space depends on being able to see.

Without vision our sensibility returns to the womb. Whether architects compose from light to dark, or from dark to light makes no difference. These are aspects of the same coin, which determines perception. To be able to see is to be aware in a way which lends significance to all of the other senses. There can be no substitute for light, for it brings into being all that we know and recognize.

Figure 7.2
Inter-Junction-City, GF-Bldg, Japan. Riken Yamamoto's dynamic use of sculptured space is emphasized by external washes of coloured ambient light. Lighting design by Riken Yamamoto & Field Shop, Japan.

8: Appendix

Nature of light

Light is a form of electromagnetic radiation - like a radio signal - transmitted in waves. Within the electromagnetic spectrum, radiations between 380 and 760 nanometres (nm) are visible to the human eye. Within this narrow band each individual colour has its own wavelength: when the whole range of wavelengths is present, the mixture of colours produced is what is described as 'white light'. But everyday terminology for describing colour is very subjective (what may be an orangey red to one person may be a reddish orange to another). The concept of light thus combines exact scientific phenomena with subjective perceptions since individual human eyes and brains behave in individual ways. Any technical understanding of light is thus continuously moderated by actual experience.

Natural light

What is thought of as natural light is a good example of this duality. Natural light is supplied principally by the sun and, to a lesser extent, by the moon and stars. The importance of sunlight or daylight is not simply that it is available and free, but that for many people daylight provides an unconscious measure of correct levels of lighting, and so, of colours. This may seem surprising: after all, daylight changes in intensity and light colour through the daily cycle, and can be altered further by weather conditions, particularly cloud and rain, as well as by the seasons. Any definition of daylight, in fact, has to be subject to conditions about the time of day and the weather: a sunny day in June has a very different quality of light to a sunny day in January. Despite this, daylight remains an unconscious constant against which all subjectively judge light. Average midday sunlight is perceived as containing all the colours in the visible spectrum, but artificial light can only approximate this, with most visible artificial light falling in the violet to red range of the spectrum in different combinations.

Colour and perception

Within the human eye, receptor cells called rods and cones interpret light falling on to the eye, and inform the sensory centres in the brain which then create the image that is seen. Rods interpret the degree of brightness, cones the degree and type of colour.

The apparent colour of an object is a function of the way in which its surface reflects light. If the surface absorbs some wavelengths (that is, colours) and reflects others, the object will appear to be in the reflected colour. Thus a red object absorbs all wavelengths except red from white light, giving a red appearance. It follows that if the light is not white, the apparent colour of the object will also change. Colours are normally defined in terms of three qualities: hue - the base tonality of red, blue, green or whatever; value - the degree of blackness or whiteness of a colour; and chroma, the degree of saturation of a colour, the purity of the colour. Tables such as the Munsell colour scale are used to define exact colours according to these three values.

Colour rendering

In lighting terms, the ability of a light source to convey colour values accurately is an important factor. This is measured by the colour temperature or colour rendering index of a lamp or luminaire. Colour temperature is calculated in degrees on the Kelvin scale: as an example, full sun at noon has a temperature of 10,000 degrees Kelvin, but of only 6,000 under an overcast sky. The colour rendering index is a more specific scale, listing lamps and fittings on a scale of 1A (best quality), 1B, 2, 3, to 4 (worst.)

Reflectance

The fact that surfaces reflect light is not only important in the perception of colour. Different surfaces reflect or absorb different quantities of light energy. This means that the redistribution of light in an interior, and so, the overall light level, is crucially affected by the reflectance of the surfaces (walls, curtains, furnishing and flooring) that make up the room.

Typical reflectance values are shown in the adjoining table, though in a critical application a designer will normally verify exact reflectance figures with the manufacturer concerned.

Artificial light sources - lamps and fittings

The majority of light sources are of course artificial, using electricity to generate light in various ways. The following list gives a general specification of each type of light-source, its colour temperature and range, and any particular advantages or drawbacks it may possess. It should be borne in mind that any lighting design is a system, that is to say it combines different elements, light sources, fittings, surfaces and colours into a whole. Achieving the whole effect is the task of the lighting designer, and the solution must be thought of as a system, not a collection of disparate elements, if the design is to succeed. There are two broad classifications of electric light-sources, incandescent and discharge, terms which relate broadly to the operating system of each. Incandescent lamps were the first to be manufactured, but new developments in fluorescent and compact fluorescent lamps have increased immensely the range of choice available.

Fluorescent lamps

Fluorescent lamps work by passing electric current through gases or metallic vapours, contained in a bulb (mercury vapour is one of the commonest used). This produces a fluorescent discharge, visible as light when it passes through the phosphor coating on the inside of the bulb or tube. At high pressures such lamps are efficient, a good part of the radiation produced by the discharge being in the form of visible light.

Standard fluorescent lamps use a single phosphor coating, which limits their colour rendering capacity. Modern lamps use a triple phosphor coating, which gives better colour rendering. The advantages of fluorescent lamps lie in their versatility, efficiency, and long lamp life. There is an excellent range of fittings now available for conventional fluorescents.

Compact fluorescents offer the performance and life of conventional fluorescents, but with integral control gear and a small lamp size. These are extremely versatile lamps for which a wide range of fittings is available.

Incandescent lamps

The incandescent lamp generates light by passing a current through a wire coil, mounted in a vacuum (or in an inert gas) until it is incandescent. Such lamps are available in mains voltage in a range of sizes and wattages. They have no requirement for control gear, and start instantaneously. A development of the GLS lamp is the tungsten lamp: this has a warm lamp colour, good for rendering skin tones. But the relatively short lamp life and high heat emission of tungsten mains lamps makes them unsuitable for large scale installations. PAR lamps have integral aluminised reflectors, allowing better control of light direction. PAR lamps have an extended life of some 2,000 hours.

Tungsten halogen lamps have better light quality than standard tungsten, with longer lamp life. They can be optically controlled more easily. The dichroic reflector lamp is a popular lamp, with an integral reflector which also dissipates much of the heat behind the lamp. New halogen lamps allow relamping with halogen into standard mains fittings, with the advantages of better rendering and modelling and longer lamp life.

Low voltage lamps

Low voltage tungsten halogen lamps are available in a large range of sizes and wattages. They offer excellent colour rendition, small lamps sizes, and long lamp life. This makes them particularly useful for display purposes. Low voltage systems require transformers to step down from mains voltage to 12 or 24 volts. The higher capital costs are normally balanced by lower running costs.

Discharge lamps

High and low pressure discharge lamps have an efficiency advantage over incandescent lamps, running cooler over a longer period. However, discharge lamps should only be used with a fixed term relamping policy, and some lamps trade higher efficiency for longer lamp life. Some lamps also only operate in fixed burning positions.

High-intensity discharge (HID) lamps use mercury or sodium as the discharge vapour. Both types require control gear, while sodium lamps are more efficient and emit an orange-white light, and mercury lamps a blueish light.

Low-pressure sodium (SOX) lamps are traditionally used in street lighting. While they are very efficient, they have virtually no colour rendering.

High-pressure sodium lamps have a long lamp life and the warm light they emit is particularly suitable for indoor areas with no daylight, such as sports halls. High-pressure mercury lamps create a cool daylight effect. They are suitable for large scale industrial areas such as warehouses. The blue cast of the light makes them unsuitable for areas where colour rendition is important.

Metal halide (MB(I)) lamps are more efficient and have good colour rendering. They are often used as substitutes for daylight. Metal halide lamps require a warm-up and restrike period.

Types of fitting

There are several general approaches to lighting that can be considered for most applications, either singly or in combination. These are downlighting, uplighting, spotlighting, wall washing, task lighting, and system lighting using either track or linear systems. There are also special requirements for environments where VDTs (visual display terminals) are in use.

Downlighting

Downlighting is understandably popular. Fitting lights into the ceiling economises on space, and is a good solution for creating a general overall light, either using recessed or surface-mounted fluorescent fittings, or recessed or suspended incandescent fittings. Glare problems can be resolved with diffusers or louvres. Spot downlighters can be used with narrow beams of light to add atmosphere.

Uplighting

Uplighting gives a soft, diffused light as most of the ambient light is reflected from the ceiling. The choice of uplighting fittings includes wall-mounted, freestanding or ceiling-suspended. Freestanding uplighters have the advantage of being easily moved, in case of reorganisation of the space, and do not require structural wiring. Fixed uplighters are less obtrusive, and can be used to enhance architectural features.

Spotlighting

Spotlights are very useful for emphasising details and thus creating visual interest. In a display setting - for example in retailing - spotlights are particularly valuable. Trackmounted or variable position spots enable a design to be revised and updated to meet new requirements. The range of lenses, reflector types, filter holders and gobos available for many spotlights offers the designer a wide choice in creating special effects.

Wall washing

Wall washing lights project an angled beam of light onto a wall surface, either from the ceiling or floor or the wall itself. This provides an attractive diffused light: as with uplighting, care needs to be taken in selecting an appropriate surface material to avoid specular glare.

Task lighting

This term describes supplementary lighting used on working surfaces such as desklights. If task lighting is to be used extensively, for example in an office, the overall ambient lighting can be reduced accordingly.

System lighting

All lighting schemes should be thought of as systems, rather than as a set of individual elements, but the term system lighting also refers to track and linear lighting arrangements in which a range of fittings performing different tasks are designed to be linked together in a coherent and adjustable whole. Track systems use a continuously powered track to which fittings can be attached at any point. Tracks can carry up to three circuits, allowing for versatile solutions with different sets of lamps on each circuit. Track can be recessed into wall or ceiling, surface mounted or suspended. Linear systems use a mixture of fluorescents and incandescents. They are highly adaptable to complex spaces, as they can be cut to size and joined either in line or at an angle. Requiring a single current feed and few suspension points, linear systems are a useful way of lighting period buildings or sites where the main surfaces need to be protected.

Lighting VDT areas

With the increasing use of computers in offices there are now strict requirements in many countries on lighting levels where VDTs are in use. These are, in general, intended to reduce operator eyestrain by avoiding glare and veiling reflections on the screen. In such contexts recessed or surface mounted fluorescent fittings are normally the best solution.

Design aspects of lighting

The question of the atmosphere of a lit space is often subjective, and a matter for discussion between designer, client and architect. But there are a number of general points to bear in mind.

Modelling is the technique of using the angle and strength of light to create relief on a surface or object. This provides visual contrast, can accentuate the mood of a space, and is a key technique in retail display and exhibition lighting. Modelling need not be over-emphatic: the drama necessary for a shop window may not be appropriate for a public space. In considering modelling, it is important to think about the surface qualities of object to be lit, to see whether to silhouette the object or emphasise its three-dimensionality. The light used as modelling light also needs to be balanced with the overall level of ambient light.

The textures of walls and ceilings can be flattened or emphasised by the placing and angle of light sources.

This technique can also add visual interest to a space: raking light across a rough-textured wall produces a pattern of shadows and sharp edges, for example.

There are various ways of highlighting a specific object (a decorative feature, for example, or a company logo). One is to use a framing head on a spotlight which will draw a clear area of light around the object. Alternatively an iris head will project a soft-edged circular area of light. A gobo can be used to project a cutout, such as a logo, in light onto a surface.

Colour systems

We have already seen that ordinary definitions of colour can be personal and subjective. There are a number of colour systems that seek to define colour more exactly, of which the Munsell system is probably the most widely used. It defines three properties of colour: hue, value and chroma. Hue is the part of the spectrum from which the colour comes (red, blue or yellow, for example). Value is related to the reflectance of the colour and is its degree of lightness or darkness. Chroma describes the intensity of the colour. Munsell set out a colour sphere, defining specific colours through a three number coordinate set.

These qualities of colour can be used by the designer. The overall colour chosen can often set the mood of a space. Strong primary colours are thought to be stimulating, cooler blues and greens are seen as restful. Such perceptions are partly psychological and partly cultural (retailers in Scandinavia prefer warm white lighting, while those in southern Europe seem to prefer cool white, for example, while particular colours are considered unlucky in some countries). Choosing different values can help distribute light in a room, for example by using a high value colour on a wall facing a window. Chroma can also be used for emphasis: a small patch of strong chroma can stimulate a design.

Colour harmony

Most spaces use a combination of colours: monochrome can be monotonous, after all. One of the best ways to find a harmonious combination of colours is by looking for differences in value and chroma: generally the larger the differences in value or chroma the more successful a combination will be. For hue there are no such clear guidelines, and this is best left to the individual taste and judgement of the designers involved.

Lighting and the environment

Since lighting is one of the most visible forms of energy consumption, it is not surprising that lighting has been drawn into the debate on energy conservation and efficiency. By definition, a good lighting design should be efficient: it should not waste energy providing more light than is needed, nor should it use inefficient systems if more efficient ones (within the whole budget of capital and running costs) can do the same task. Reducing light levels to save energy (given the small proportion of energy used by lighting in a modern office building or mall) will be a false economy if it reduces staff effectiveness or safety.

Energy efficient choices include the increasing range of compact fluorescent lights, the new 26 mm diameter fluorescent tubes, and new mercury free lamps. The choice of reflectors and fittings to deliver the maximum level of light to the right area is also important, and photometric data, which analyses the levels and directions of light output from a fitting, is the starting point for such an analysis (photometric diagrams are part of the presentations of fittings in this program). Maintenance is another important factor in maximising lighting efficiency.

Control systems

Control systems are a major way of using light efficiently and so making energy savings. The simplest systems, dimmers, allow light levels to be reduced when the maximum is not required. Linking light switches to sensors in hotel and office corridors ensures that light is only on when it is needed, for example, and similar systems can balance daylight and artificial light in offices or compensate for overlighting in retail window displays at night.

CIE guidelines

The CIE guidelines for energy conservation also form an excellent checklist for any lighting design. They are as follows:

- Analyse the difficulty, duration, criticalness and location of the task in order to determine lighting needs, and give consideration to the visual differences between people.
- Ensure that the design scheme complies with current regulations and recommendations regarding light levels.
- Choose the most efficient and appropriate lamps.
- Select fittings which are efficient and give a correct light distribution without glare or veiling.
- Use the highest practical room surface reflectances to maximise lighting efficiency. Integrate lighting with heating and air conditioning systems to save energy.
- Ensure that flexibility in the system allows sections to be turned off and lighting levels to be reduced as needed.
- If appropriate, incorporate daylight into the artificial lighting system, while ensuring itdoes not cause glare or an imbalance of brightness.
- Establish a regular, well-informed level of maintenance for cleaning luminaires and room surfaces, and for relamping.

Glossary

Apparent colour: Of a light source; subjectively the hue of the source or of a white surface illuminated by the source; the degree of warmth associated with the source colour. Lamps of low correlated colour temperatures are usually described as having a warm apparent colour, and lamps of high correlated colour temperature as having a cold apparent colour.

Average Illuminance (Eave): The arithmetic mean illuminance over the specified surface.

Brightness: The subjective response to luminance in the field of view dependent upon the adaptation of the eye.

Candela (cd): The SI unit of luminous intensity, equal to one lumen per steradian.

Chroma: In the Munsell system, an index of saturation of colour ranging from 0 for neutral grey to 10 or over for strong colours. A low chroma implies a pastel shade.

Colour constancy: The condition resulting from the process of chromatic adaptation whereby the colour of objects is not perceived to change greatly under a wide range of lighting conditions both in terms of colour quality and luminance.

Colour rendering: A general expression for the appearance of surface colours when illuminated by light from a given source compared, consciously or unconsciously, with their appearance under light from some reference source. Good colour rendering implies similarity of appearance to that under an acceptable light source, such as daylight. Typical areas requiring good or excellent colour rendering are qualify control areas and laboratories where colour evaluation takes place.

Colour rendering index (CRI): A measure of the degree to which the colours of surfaces illuminated by a given light source conform to those of the same surfaces under a reference illuminant.

Colour temperature (Tc, unit: K): The temperature of a full radiator which emits radiation of the same chromaticity as the radiator being considered.

Contrast: A term that is used subjectively and objectively. Subjectively it describes the difference in appearance of two parts of a visual field seen simultaneously or successively. The difference may be one of brightness or colour or both. Objectively, the term expresses the luminance difference between the two parts of the field.

Correlated colour temperature (CCT, Tcp, unit: K): The temperature of a full radiator which emits radiation having a chromaticity nearest to that of the light source being considered, e.g., the colour of a full radiator at 3500 K is the nearest match to that of a white tubular fluorescent lamp.

Direct ratio (DR): The proportion of the total downward luminous flux from a conventional installation of luminaires which is directly incident on the working plane.

Effective reflectance: Estimated reflectance of a surface, based on the relative areas and the reflectances of the materials forming the surface.

Flux fraction: The proportion of luminous flux emitted from a luminaire in the upper or lower hemisphere (upper and lower flux fractions).

Flux fraction ratio (FFR): The ratio of the upward luminous flux to the downward luminous flux from a luminaire. It is also the ratio of the upper flux fraction to the lower flux fraction and the ratio of the upward light output ratio to the downward light output ratio.

Full radiator: A thermal radiator having the maximum possible radiant excitance for all wavelengths for a given temperature; also called a black body to emphasise its absorption of all incident radiation.

Glare: The discomfort or impairment of vision experienced when parts of the visual field are excessively bright in relation to the general surroundings.

Hue: Colour in the sense of red, or yellow or green etc. (see also Munsell)

Ingress protection (IP) number: A two-digit number associated with a luminaire. The first digit classifies the degree of protection the luminaire provides against the ingress of solid foreign bodies. The second digit classifies the degree of protection the luminaire provides against the ingress of moisture.

Illuminance (E, units: lm/m2, lux): The luminous flux density at a surface, i.e. the luminous flux incident per unit area. This quantity was formerly known as the illumination value or illumination level.

Initial light output (unit: lm): The luminous flux from a new lamp. In the case of discharge lamps this is usually the output after 100 hours of operation.

Installed power density (W/m2/100 lux): The installed

power density per 100 lux is the power needed per square metre of floor area to achieve 100 lux on a horizontal plane with general lighting.

Irradiance (Ee, E, unit: W/m2): The radiant flux density at a surface, i.e. the radiant flux incident per unit area of the surface.

Light output ratio (LOR): The ratio of the total light output of a luminaire under stated practical conditions to that of the lamp or lamps under reference conditions. For the luminaire, the output is usually measured in the designated operating position at 25 Cambient temperature with control gear of the type usually supplied in a luminaire and operated at its normal voltage. For the lamp the output is measured at 25 C ambient temperature and with control gear of standard properties. This is a practical basis forevaluating the total light output to be expected under service conditions.

Lumen (lm): The SI unit of luminous flux, used in describing a quantity of light emitted by a source or received by a surface. A small source which has a uniform luminous intensity of one candela emits a total of 4 x pi lumens in all directions and emits one lumen within a unit solid angle, i.e. 1 steradian.

Luminaire: An apparatus which controls the distribution of light given by a lamp or lamps and which includes all the components necessary for fixing and protecting the lamps and for connecting them to the supply circuit.

Luminance (L, unit: cd/m2): The physical measure of the stimulus which produces the sensation of brightness measured by the luminous intensity of the light emitted or reflected in a given direction from a surface element, divided by the projected area of the element in the same direction. The SI unit of luminance is the candela per square metre.

Luminous flux (unit: lm): The light emitted by a source, or received by a surface. The quantity is derived from radiant flux by evaluating the radiation in accordance with the spectral sensitivity of the standard eye as described by the CIE standard photometric observer.

Lux (lux): The SI unit of illuminance, equal to one lumen per square metre (lm/m2).

Maintenance factor (MF): The ratio of the illuminance provided by an installation at some stated time, with respect to the initial illuminance, i.e. that after 100 hours of operation. The maintenance factor is the product of the lamp lumen maintenance factor, the lamp survival factor (where group lamp replacement without spot replacement is carried out), the luminaire maintenance factor and the room surface maintenance factor.

Mounting height (hm): Usually the vertical distance between a luminaire and the working plane. In some cases the floor may be the effective working plane.

Purity: A measure of the proportions of the amounts of the monochromatic and specified a chromatic light stimuli that, when additively mixed, match the colour stimulus. The proportions can be measured in different ways yielding either colorimetric purity or excitation purity.

Radiance (unit: W/m2/sr): At a point on a surface, the quotient of the radiant intensity emitted from an element of the surface in a given direction divided by the projected area of the element in the same direction.

Reflectance (factor) (R, p): The ratio of the luminous flux reflected from a surface to the luminous flux incident on it. Except for matt surfaces, reflectance depends on how the surface is illuminated but especially on the direction of the incident light and its spectral distribution. The value is always less than unity and is expressed as either a decimal or as a percentage.

Saturation: The subjective estimate of the amount of pure chromatic colour present in a sample, judged in proportion to its brightness.

Utilisation factor (UF): The proportion of the luminous flux emitted by the lamps which reaches the working plane.

Value: In the Munsell system, an index of the lightness of a surface ranging from 0 (black) to 10 (white). Approximately related to percentage reflectance by the relationship $R = V(V-1)$ where R is reflectance (%) and V is value.

Weight: An approximate correlation of Munsell value modified to give, in conjunction with greyness, subjective equality of brightness in the various hues.

Working plane: The horizontal, vertical, or inclined plane in which the visual task lies. If no information is available, the working plane may be considered to be horizontal and at 0.8 m above the floor.

The explanations and definitions given in this Glossary are based on BS 4727: Part 4: Glossary of terms particular to lighting and colour (1971/1972) (2 4) and on the International Lighting Vocabulary 1987 (Il2) issued jointly by the Commission Internationale de l'Eclairage (CIE) and the International Electrotechnical Commission (IEC). These should be consulted if more precise definitions are required. The following values are typical reflectances for the finishes stated. If an exact figure is required for the reflectance of a specific material, such as a paint or wallpaper, the manufacturer involved should be consulted

Finish	Reflectance	Finish	Reflectance
White emulsion paint of plain plaster surfaces	0.8	Birch or maple panelling	0.35
White glazed tiles	0.8	London stock brick	0.25
White paper	0.8	Light oak or mahogany panelling	0.25
White paint on acoustic tiles	0.7	Light grey concrete	0.25
White glazed brick	0.7	Marbled PVC tiles	0.25
Pink plaster	0.65	Medium oak or teak panelling	0.2
White emulsion paint on concrete	0.6	Dark grey brick	0.2
White emulsion paint on wood	0.5	Polished cork tiles	0.2
Cement screed	0.45	Blue engineering brick	0.15
Off-white PVC tiles	0.45	Quarry tiles	0.1
Light grey or pale carpet	0.45	Dark carpet	0.1
Concrete brick	0.4	Dark oak panelling	0.1
Portland or white cement	0.4	Black chalkboard	0.15
Stainless steel	0.35		

Bibliography

Boyce, P R, *Human Factors in Lighting*, Conran Octopus, 1985
CIBSE, *Lighting Guide to the Outdoor Environment*, 1992
Duncan, Alastair, *Art Nouveau Lighting, Thames and Hudson*, 1978
Duncan, Paul, *Lighten Our Darkness*, Royal Fine Art Commission
Durrant, D W, *Daytime Lighting in Buildings*, IES
Gardener, Carl and Hannaford, *Lighting Design*, Design Council, 1993
Lam, William, *Perception and Lighting as Formgivers for Architecture*, McGraw-Hill. 1977
Lynes, J A, *Principles of Natural Lighting*, Elsevier
Moyer, J L, *The Landscape Lighting Book*, Wiley, 1992
Myerson, Jeremy, *Better Lighting*, Conran Octopus, 1985
Phillips, Derek, *Lighting in Architectural Design*, McGraw-Hill
Turner, J, *Lighting*, B T Batsford Ltd, London, 1995
Wells, Stanley, *Period Lighting*, Pelham Books, 1975

Index

accent 36-7, 78, 100, 103, 105
ambient 16, 36, 98
art galleries 58-73
art nouveau 93
atria 21, 23, 24, 38
audiovisual 69

bathrooms 46, 108
bedrooms 50, 107
Bernini 18, 21
Bing, Sam 93

candles 52, 88
chandeliers 46, 104
Chartres 7, 8
colour rendition 36, 41, 62, 64, 65
conferences 42, 54
conservation 124
control 14, 50-51

dining rooms 104
display 60-68, 101
downlights 40, 50, 71, 100, 122

Einstein 110
entrance 40, 46, 96
exterior 40, 49, 76

fabric structures 28
facades 82-5
fibre optics 31, 32, 38, 39, 46, 50, 52, 63, 82, 88
filters 21, 32
floods 21, 82
fluorescent 18, 31, 39, 46, 121
futurists 115

gardens 77-80
glass blocks 26
gobos 13, 108
Gray, Eileen 94
Gropius, Walter 8, 110

Hagia, Sofia 4
hotels 17, 24, 40-57

Japan 117

Kettle's Yard 86
kitchens 106

lanterns 80
Las Vegas 48
lightpipes 22-3
lightshelves 26
living rooms 102-3
louvres 21, 35, 55, 63
low voltage 31, 32, 37, 88, 100, 121

malls 16, 17, 30-39, 38
merchandise 16, 30, 34, 36-7
mercury vapour 80, 121, 122
metal halide 22, 32, 46, 52, 54, 76, 82

neon 8, 82

offices 10, 22-3, 26, 39
Olesen, Otto K 74

PIR detectors 76
plants 22, 24, 38, 56, 78
pole lanterns 47, 85
presentation 14

radiation 63, 121
reflectance 26, 55, 63
restaurants 52, 70
residential 90-108
van der Rohe, Mies 8, 104, 112

Saint Denis 4
sky pollution 49, 74
skylights 23, 26
solar energy 28
spotlights 31, 34, 36, 37, 63, 10, 105, 107, 122
stake luminaire 76
starpoints 32
swimming pools 41, 55-6, 82, 98, 103

task lamps 98, 103, 122
Tiffany 93
tungsten halogen lamps 31, 34, 37, 38, 56, 63, 76, 121

Urban, Joseph 114

wallwashers 40, 54, 122
Wright, Frank Lloyd 8, 18, 94

zen 77, 86